PASTURED POULTRY PROFITS

by

Joel Salatin

Polyface, Inc.
Swoope, Virginia

Pastured Poultry Profits, First Edition
Copyright © 1993 by Joel Salatin

First reprinting 1996
Second reprinting 1999

Photographs by Teresa W. Salatin

Editing and Layout by Vicki H. Dunaway

Cover design by Keystrokes and Design

Library of Congress Catalog Card Number: 93-86302
ISBN: 0-9638109-0-1

Contents

WHY PASTURED POULTRY?

IN THE BEGINNING

OUT TO PASTURE

PROCESSING

PROBLEMS (IT'S NOT ALL ROSES)

MARKETING

POSSIBILITIES

APPENDICES

Foreword

March 19, 1991

In this book a proven production model is described, which is capable of producing an income from a small acreage equal or superior to that of most off-farm jobs. However, this production model is capable of producing more than money. It is capable of producing a very high quality of life as anyone who has ever visited Joel and Teresa's farm can attest.

As much as I admire Joel Salatin's agricultural engineering abilities, I admire even more his steadfast refusal to compromise his family's quality of life. Probably the greatest lesson we can learn from Joel's success is that if we produce a high enough quality product the marketplace will bend and adapt itself to meet our needs, wants and desires. We just need to have faith and stand our ground.

This production model should be viewed as a whole. All of the recommendations contained in this book are the results of many years of costly trial and error and should not be dismissed lightly in your initial efforts. This model will not return the same dollars if it is done on a non-seasonal basis, or if the chickens are sold in any method other than live and direct. Most of the disappointments with pastured poultry have come from people who thought they could come up with an "easier" production or marketing model than Joel is using.

As Joel points out, you should start off producing no more chickens than you and your family can consume until you are sure you can produce a superior product. The first person you have to sell on your chicken is you. Your best advertising will be the exceptional taste of your chicken. Give some to your friends and neighbors and let your business grow slowly from word of mouth referrals. Start small and give yourself plenty of time to see if pastured poultry truly fits your needs and life-style before investing a lot of capital.

Due to the necessity to condition the pasture for the chickens and to rotate the land where the chickens are grazed, this production model is best used in combination with some form of ruminant livestock production. In addition to poultry, Joel also sells excellent grass-fed beef.

All of us interested in seeing an American rural revival should thank Joel and Teresa Salatin for going to the effort and expense to share their experiences and recommendations for pastured poultry with you.

Allan Nation
Editor, *The Stockman Grass Farmer*

Introduction

A couple working six months per year for 50 hours per week on 20 acres can net $25,000-$30,000 per year with an investment equivalent to the price of one new medium-sized tractor.

Seldom has agriculture held out such a plum. In a day when main-line farm experts predict the continued demise of the family farm, the pastured poultry opportunity shines like a beacon in the night, guiding the way to a brighter future.

I've pastured poultry in portable pens since the late 1960's, when I was scarcely a teenager. It worked then, and it works now.

Because so many people have sought information on our model, I have attempted to put it in writing so that it may be duplicated throughout the world. For the health of the rural economy, the health of the environment, the health of the poultry consuming public, and the health of farm families, we offer this proposal.

We encourage continued refinements and innovation to this system, and trust that our role as catalyst will encourage thousands of enterprising farm families.

About Polyface Farm

Polyface, Inc. is a closely held incorporated family farm established by my parents, William T. and Lucille Salatin. Located just southwest of Staunton, Virginia, in the scenic Shenandoah Valley of Virginia, the farm is currently operated by my mother and my family: wife, Teresa Wenger Salatin; son, Daniel and daughter, Rachel.

My grandfather, Frederick Salatin, was a charter subscriber to *Organic Gardening and Farming* magazine, and was a master gardener and craftsman from Indiana. Dad and Mom moved to Virginia in 1961 and carried the organic farming torch from Grandpa. I am the privileged third generation to continue the principles based on the belief that God created the Earth and established humanity as its steward, to nurture, protect and embellish. This philosophy precludes the use of toxic chemicals, debasing substances, and erosive practices, and instills instead an insatiable thirst for agricultural truth.

The truth manifests itself in natural principles of plant and animal life. The farm should capitalize on these laws rather than fight against them.

Roughly 550 acres, the farm is about 100 acres open land and the balance wooded. We produce organic grass-fattened beef, home-grown broilers, firewood, eggs, rabbits and vegetables. Full-time farmers who are not independently wealthy, we rely on the farm for our sustenance and have developed production, processing and marketing systems that make an end run around conventional agriculture's roadblocks.

We can hardly imagine being more enthusiastic about farming, and trust that these pages will encourage and inspire farmers of all ages to go forward with renewed optimism.

Joel Salatin
July 1993

WHY PASTURED POULTRY?

Chapter 1

The Pastured Poultry Opportunity

A couple working 50 hours a week for six months a year on 20 acres can NET $25,000-$30,000. What other agricultural enterprise can do that?

Read that first paragraph again. Let it sink in. There are certain times in the history of any culture when golden opportunities go begging. This is one of those times. All the indicators point to pastured, home processed poultry as one of this century's best family farm enterprises.

Per capita consumption of red meats is down while poultry and fish are on a dramatic incline. Who would have predicted in 1900 that poultry would surpass beef in per capita consumption? It has. And the lines on the chart continue their divergent paths. Americans are buying chicken.

But there is a catch. The poultry industry's factory farming methodology, its automated processing plants with concomitant *Salmonella* and sickness outbreaks among consumers have completely adulterated the product. Media attention on the vertically integrated poultry industry is unprecedented. National TV news programs and magazine articles, and a crusade from animal rights activists against inhumane confinement, have left many consumers rightfully concerned

2

about the safety and environmental costs of poultry products in the supermarket.

If there were an alternative - some cleanly, humanely produced poultry -consumers would flock to it, no pun intended. Many consumers have turned vegetarian in the face of perceived filth and animal cruelty. Justified or not, this reaction further contracts the pool of consumers who otherwise would patronize the meat counter.

Nutritional perceptions, coupled with new environmental sensitivities, are creating a generation clamoring for natural, organic, eco-logical, biological, regenerative food. Call it what you will, everyone knows deep down what these ideas mean, in spite of ongoing academic debates over definitions.

The fact is that many consumers want to exit conventional food channels, some of necessity, some of conviction, some of mistrust, and some simply because somewhere they've tasted clean food and found it memorable. Pastured poultry offers an alternative to all of these consumers.

Pastured chickens, because they eat high amounts of forage, can be clinically shown to be far lower in saturated fat than conventionally produced birds. Raised without the negatives - antibiotics, steroids, fecal air and artificial light - but with the positives - probiotics, kelp meal, natural vitamins, fresh air and sunshine, clean pasture paddocks and in small groups - pastured birds offer a completely different meat to the consumer.

Pasturing allows these birds to be grown without damaging substances. The industry says Cornish Cross broilers cannot be raised without artificial vitamin packs and antibiotics. They are right, if the birds are raised in a setting as unnatural as factory confinement houses. But out on pasture, with fresh air, sunshine, green material and wholesome feed, these broilers will outperform their factory counterparts in every way. They will gain better, convert feed more efficiently, be healthier (fewer culls for any reason, including benign tumors) and possess a superior taste. That makes them easy to sell and easy to eat. It allows competitive production costs, all the while producing a more nutritious, clean product.

Each bird requires about 5.5 person minutes during the production phase and about 3.5 minutes during the processing phase. Each chicken, therefore, allowing for slippage, requires roughly 9 person minutes. Fixed costs run approximately $1.80 - $2.00 per bird. At a selling price of $1.35 per pound and an average carcass weight of 4 pounds, this yields about $3 profit per bird, or $12-$20 per hour for the producers' time.

Machinery investment runs between $10,000 and $15,000, but that can be expended over time and can be reduced by procuring second hand equipment.

For six months, the couple works hard producing 10,000 broilers, processing two long days a week for 20 weeks. Daily chores on the five nonprocessing days take about 4 hours. The other six months can be spent loafing, working or whatever.

Many things can make this model NOT work. Trying to shortcut the pasture element, misman-agement, hiring the processing done off the farm, shortcutting ration ingredients, or a host of other things. Success is not guaranteed. People are not equal in their desire or ability to

4

succeed as entrepreneurs. Just because you raise a good product, or just because you have a novel approach, you are not free from the rules of prudent business.

But the potential is there. It has been done and it can be done. That is for sure.

This opportunity applies to a broad spectrum of people. Lest you miss the personal application, I will describe where this opportunity fits. It is certainly a way for a couple to get into farming. Whether you own land or not, it is a way to start farming and make it a full-time vocation.

A family currently farming, but floundering for a way to make it pay, is a prime candidate. Perhaps you are seeking a way to turn your financially distressed farm around. Here is a way.

Perhaps you are an elderly farmer and you have children who want to farm. You have discouraged them because everyone knows "there's no money in it." So, along with your peers, you have encouraged your children to leave the farm and spend their lives doing something more financially rewarding. Perhaps it is too late for you to begin this enterprise, but you can have the vicarious thrill of watching or helping your children build a financially and emotionally rewarding farming future.

Or perhaps you have fished around for something to occupy a son or daughter on the farm. Your farm may be making a living for you, but you don't see how it can provide employment for your child and family too. Your heart's desire is to work alongside your child in your older years, to enjoy the camaraderie and the satisfaction of watching the farm pass on to the next generation. Here is a way for that next generation to earn a healthy living without impacting the rest of the farm, without taking away any income from the normal farm operations.

Maybe you are a second or third child. Dad and Mom set up the firstborn on the farm, but now you're coming along and you'd like to farm too, but the farm just can't pay another salary. Ask for a corner somewhere and show that you can earn your own way. They may poke fun at you at first, but you can cry all the way to the bank.

Perhaps you are a consumer who has no desire to farm, but you would like healthy food and you know an enterprising young person who would be perfect for this. Pass this opportunity along. In short, this opportunity spans America: urban and rural, young and old, male and female. Let the dream belong to you or someone you know.

Just remember, a couple working 50 hours a week for 6 months a year on 20 acres can NET $25,000-$30,000. That's better than a city job.

Chapter 2

What is Wrong With
the Poultry Industry?

Many people from the alternative foods movement do not think a radical overhaul of the vertically integrated poultry industry is necessary to produce a chicken worth eating. They want to minimize the changes, thinking that a few changes here and there will make good chicken.

I think it is important to point out just how far away from "correct" the mainline poultry industry is. When we examine it all in one place, I think the evidence will indeed suggest that the whole mess is rotten, beyond salvage, and ought to be thrown out.

First, of all, the poultry capitals are not located near the grain capitals. This means that the poultry centers must import their grain from hundreds of miles away, and the grain centers lose the manure nutrients generated from their grain. The obvious result is a nutrient overload in the poultry production areas and a nutrient shortfall in the grain producing areas.

This nutrient overload generally goes into the groundwater, surface water runoff and into the air as ammonia vapor. The soil can only assimilate a certain volume of nutrients. This is why environmental groups and rural house-and-lot folks are passing ordinances about poultry houses. Farmers say their property rights are being infringed upon, that they ought to be able to build what they want where they want when they want, and they should be able to spread that manure when, where and how they want.

The polarization is enormous. The tragedy is that it doesn't have to be that way. Our model does not smell, does not pollute, angers no neighbors with noxious odors, and kills no babies through nitrate-induced Sudden Infant Death Syndrome (SIDS). The point is that these problems are inherent in a system that concentrates poultry production, that worships at the altar of a fast buck instead of long-term societal well-being.

Grower concentration is necessary because a processing plant can only service growers within a 100-mile radius or so. As soon as you concentrate processing, production must be concentrated as well. The size of the processing facility and the concentration of poultry inherently cause many of the problems. We can't just take the current industry, feed the birds organic feed and natural supplements, and have an acceptable system. The fact is that we must decentralize the production, thereby spreading the manure around in a light enough application that the soils can assimilate the nutrients.

The next major flaw is the whole notion of confinement housing, or "factory farming," as the Humane Society calls it. The inherent fecal contamination in such a model causes all sorts of health problems. Feathers, eyes, beaks, nostrils - nothing is exempt from a layer of fecal dust and its pathogen-laden microorganisms. The birds breathe in the fecal dust, which contains a high percentage of ammonia, and this causes lesions of the respiratory linings, the fragile mucous membranes.

Perhaps no single problem in the industry receives more attention than this. That is why the chicken industry has now gone to nipple waterers that resemble a guinea pig waterer. It's not because it is easier, but because the industry is trying to get away from fecal contamination that coats water pans, feeders, and everything in a house. Fecal dust is everywhere, and the birds live in that their whole lives.

Fecal contamination, and certainly ammonia, can both be diminished by a proper carbon-nitrogen ratio in the bedding, and a composting floor litter. But even with the closest attention to detail, a house of 10,000 birds is still going to have a major problem. It's just too many birds in too tight a space. Air exchange and management simply can't keep up with such a volume.

Because the factory house has such inherent problems, productivity is maintained by feeding antibiotics and hormones, poisons to enhance the appetite (like arsenic), heavy metals and a host of other additives that increase meat toxins. The meat therefore becomes soft, water-absorbent, lacks muscle tone and is violently toxic to environmentally sensitive people. This environment stimulates drug-resistant *Salmonella*, which the human body was not made to handle. The medical community is publishing startling discoveries about R-factor disease strains. Dubbed "superbugs" by researchers, these organisms are resistant to traditional antibiotics and have come into existence as a direct result of drug feeding on farms.

Now we move to the processing, which begins after birds are loaded on trucks and driven for

9

up to three hours to the processing plant. Mechanical killing requires a perfectly still bird so the cutting wheel will hit the jugular vein every time. When the birds hang upside down on the shackles, they twist and turn - not conducive to mechanical killings. To get them still, an electrical current stuns them. This keeps the birds from bleeding well and accounts for much of the black clotted blood around the bones of conventional birds.

Mechanical evisceration breaks open intestines and pours fecal material over the carcass, inside the body cavity, and contaminates the birds. Large chill tanks often have several inches of fecal sludge in the bottom. In fact, about 9 percent of the weight on department-store chicken is fecal soup. The soft muscle tissue is more conducive to insoaking, and the carcass sponges up the fecal-contaminated chill water. Of course, this adds to the carcass weight, but certainly does not contribute any to the health of consumers.

This filth is why birds receive as many as 40 chlorine baths - how much of that permeates the meat? And now the Food and Drug Administration has approved irradiation of chicken to control *Salmonella* and other bacteria that are a direct result of high-speed automated processing. Irradiation reduces vitamin C levels and reduces the nutrients in the meat. Processing is an inherently filthy thing. And the larger, the faster, the more automated the system, the more filthy it is.

And all we've talked about is the chicken. We have not discussed the farmers who have been reduced to feudal serfs, subservient to the vertical integrator. Many farmers are jerked around by the conglomerates that promise them the world in a factory house. Processing plant workers, mere cogs in a wheel, are demeaned to a work situation that often is neither personally rewarding nor physically healthy. Carpal tunnel syndrome, known generically as repetitive motion disease, afflicts many processing plant workers. The bottom line is that as the industry grows and spits out chickens, it uses and spits out growers and workers.

To be sure, the industry and its regulatory cronies deny these things - all of them - but that is typical establishment protectionism. Look how the USDA touted DDT as the best thing since toilet paper before Rachel Carson wrote *Silent Spring*. Truth denial by interests who stand to gain by misleading people is a common strategy the world over. Spanish has an idiom for this: "Tell them any old silly." When the peasants, the public, begin questioning, then the mainstream academic, government and big business communities join to tell people "any old silly." Another person placated, another dollar in their pockets.

While I am not a Henny Penny screaming that the sky is falling, neither do I believe many of the statements made by the powers that be. The stakes are too high for bigwheels to level with people. My years as an investigative news reporter showed me how much powerful people think they can get by with. And not until evil people, evil empires, or evil agricultural practices are exposed to the public and brought to the citizen's bar of justice (the voting booth or market dollar) do the facts ever begin to emerge.

That is why I do not think we can change the giants. Forget them. Don't fret about them. Let them go ahead and do whatever they want to do.

Just grow good chickens, one at a time, give some local folks a healthy alternative, convert the vegetarians, and devote your energies to a positive alternative for your family and your community. That is far more rewarding than a negative campaign against the poultry giants.

Where does this conventional production leave us? We end up with a chicken that has been raised in a horrible way, fed horrible feed, processed horribly and isn't fit to eat. The entire establishment does not deserve consumer patronage. It is not a pretty picture, but it explains the fuel being added to the animal rights movement, the environmental movement and the anti-farmer sentiment in the public policy arena. It all stems from inherently improper models in the poultry industry. And it needs to be scrapped. It is fundamentally flawed, and the sooner we realize it and start over with radical restructuring, the better for the farmer, the environment and the consumer. Indeed, the better for our culture's collective health.

Chapter 3

Fat Animals, Fat People

"**Y**ou are what you eat." This trite old adage, commonly accepted, should challenge forage producers willing to crack the retail market niche of meat, milk and eggs.

Chlorophyll, that wonderful material that makes photosynthesis happen and inhabits all green plants, is a detoxifier. Health food stores for decades have carried chlorophyll capsules as a natural Roto-Rooter to clean out and detoxify the body.

Magnesium, a large component in chlorophyll, is equally important, as are the B vitamins, linked by naturalists to everything from insomnia to nervousness to epilepsy. Whole books using both replicated and anecdotal research illustrate the validity of the nutritionists' claims.

When mothers sternly tell their children to eat their greens, they aren't forcing the youngsters to take poison. They, as well as our salad society, know the value of green material in the diet.

Green material is high in vitamins and

minerals, but low in energy. In fact, humorists point out that it takes more calories to chew up and swallow lettuce than the green vegetable offers in return. So why eat it? For the chlorophyll, the vitamins, the minerals and the fiber.

The shift away from high energy foods in our culture reflects our change from heavy, outside manual labor to the stressful, fast-paced, sedentary lifestyle of most Americans.

Consumers say they don't want the saturated fat and high energy of animal proteins. The American Heart Association, doctors and even components of the USDA warn people about cholesterol levels in animal proteins, encouraging people to use vegetable oils and salads instead.

As a percentage of total calories per pound of material, greens are low. In comparison, grains and especially corn, are high in energy but low in vitamins and minerals.

When animals are fed a high energy, low vitamin/mineral diet, they tend to have more saturated fat (cholesterol) in their meat, milk or eggs, just like people. When these same animals consume a large percentage of green material, the saturated fat of their animal proteins diminishes.

If you have a cholesterol problem, your doctor will tell you to do several things: cut down on animal proteins (meat, eggs and dairy products), exercise in order to metabolize the excess calories in your body (isn't that a euphemistic way of saying "fat"?), eat salad for lunch instead of meat and potatoes. A good doctor will also tell you to identify and then reduce stress in your life. The stress-cholesterol connection is clear.

There you have it: diet shift, exercise, stress reduction. Those are the key elements.

Now let's look at American agriculture, which for decades has crowded animals into confinement feeding and housing facilities, causing tremendous stress. In such tight housing, exercise is practically nonexistent, epitomized by caged laying hens and crated veal calves. And to top it off, their diet has steadily shifted from an outside pasturage component with supplemental grain feeding to extremely high energy rations devoid of green material. Current mainline livestock agriculture ensures the very elements that encourage saturated fat: high calorie/low vitamin-mineral rations, sedentary lifestyle and both physical and emotional stress.

Is it any wonder that cardiovascular disease and the cholesterol problem follow proportionately?

Yellow fat on poultry and beef, extremely orange egg yolks, and naturally yellow butter reflect high levels of chlorophyll in the diet and low levels of saturated fat. In fact, some nutritional doctors recommend these types of eggs to help detoxify their patients. Cholesterol isn't a problem. We've even seen folks reduce cholesterol by eating fresh forage-based eggs.

We first became aware of this animal diet/ production model and fat connection when the dietician at our local hospital analyzed one of our broilers and one out of the store. She cooked them both the same amount of time and made "fat" comparisons. Ours was completely different and she immediately began recommending our birds to heart patients.

The first time a professional barbecued several hundred halves of our chicken for a field day, he was astounded at the difference. He said flatly, "Look, chicken is chicken." But then he made a startling discovery. He said that carcass weight normally drops 20 percent during the cooking process. Ours lost only 9 percent. In all the thousands of birds he'd cooked, he'd never experienced anything like it.

Interestingly, research suggests that conventional chicken, because it lacks muscle tone (the meat is soft and mushy rather than firm and solid) insoaks a substantial amount of water from the chill tanks in a commercial processing plant. In fact, a figure bandied about is that up to 10 percent of the retail meat counter weight of conventional chicken is water from the chill tanks. If that is true, it is certainly coincidentally close to the 11 percent difference between the cooking loss on our birds and conventional ones. If it's not true, the difference simply substantiates the "grease drippings" difference and lends more credence to the notion that altering production models can completely alter the quality of the meat, milk or eggs.

Chill tanks often have a layer of fecal sludge on the bottom because of the volume of carcasses and the bits and pieces of excrement that were not washed off during processing. This water insoak, therefore, has an accompanying fecal contamination level. Is it any wonder the poultry industry wants to irradiate poultry? Furthermore, the industry is fighting proposals to mandate air chilling instead of water chilling. If the 10 percent figure is accurate, it's easy to see why the industry opposes air chilling: the retail weights would drop significantly.

We proponents of forage-based agriculture must be aggressive about differentiating between animal proteins produced with greens as the main dietary component and those produced without greens. That these differences exist is fact.

Our range raised broilers, for example, have yellow fat, and when cooked do not make hard grease on top of the broth. It stays softer than Crisco. On conventional birds it hardens like wax.

Hot weight to cold weight shrinkage in beef is a major problem in the beef industry. The normal range is 2-4 percent. In all the years we've been producing forage fattened (salad bar) beef, we've never lost one ounce to shrinkage, even when the carcass has hung for two weeks. The protein is pure food, not so much soft energy in white fat.

It is time for confinement dairies, feed-lots, broiler houses and egg factories to defend themselves in a court of ethics. Their practices not only foster a row crop agriculture and its accompanying use of chemicals and topsoil loss, but engenders in our society needless sickness. And for those of us who dare to say animals, just like people, differ according to their production and processing, let the record show that greens reign supreme. Grass fed animal proteins offer consumers a vicarious salad - good food without guilt. Fat animals and fat people go together.

Chapter 4

Family Background

In 1967, when I was 10 years old, I purchased 50 straight-run heavy breed chicks from Sears and Roebuck, beginning a love affair with farming and business that gets stronger each day.

As usual, two-thirds of the chicks turned out to be roosters. The 18 hens supplied eggs for the family and enough to sell. Dad and Mom bought eggs from me, and I peddled the balance to folks at church and in the community.

We dressed the roosters, heating the water in a big pot over a fire and plucking the birds on the clothesline in the backyard. Dad showed me how to gut a chicken and we were off and running.

About that time, my older brother, Art, was developing a rabbit business. Realizing the importance of low-cost production models and the benefits of moving livestock around, Dad immediately set to work designing a portable rabbit rearing system.

For some reason, we visited a man about 20 miles away who had portable hutches in his pasture, and Dad was enamored with the idea of moving the buildings around in the field. I still can't remember what the man had in the hutches, but they were A-frames. I think he had pigs in there, but it doesn't matter now. Of course, Dad was no doubt familiar with portable chicken houses on skids that farmers would pull around their fields with horses. Any poultry book published before 1950 has photographs of such structures.

Anyway, Dad had already built a portable veal calf barn with slatted floor and four quadrants. The milk cows would shade up near it and graze next to it in the pasture. At milking time, the calves would jump outside, suckle the nurse cow tied to a corner of the veal trailer, and then go back in for some fresh hay and grain. This procedure kept the calves close to the cows, spread the droppings in the pasture, and gave the cows a comfortable place to lounge.

Dad was very much a visionary and inventor, developing a portable electric fence system in the early 1960s. Dad immediately set to work designing a portable rabbit facility. He came up with a squatty rectangular house containing four quadrants for four does with a narrow door in each that opened into an enclosure. The enclosures mounted against the house like wings off an airplane. The wings were 8 ft. x 12 ft. long and divided lengthwise down the middle so that each quadrant's rabbits (normally either a doe or a doe and her litter of bunnies) had access to a run that was 4 ft. x 12 ft.

By moving the whole contraption every few days, we could provide the rabbits with fresh forage, keep them much cleaner and healthier, and dramatically reduce feed costs. After all, rabbits are herbivores. But, alas, rabbits like to dig.

And this was our Waterloo. We tried many things, including poultry netting buffers on the edges. Art even drove 1 inch nails in the bottom board every inch or so sticking down. But no matter what we tried, we could not keep the rabbits in. Any little depression in the ground was an invitation to escape. We spent many hours rigging up string, stick and box traps, and hiding behind the corner of the shed, waiting to pull the string and catch rabbit escapees. We probably spent more time chasing rabbits than actually raising rabbits.

Finally, we abandoned the idea for rabbits and began carrying them fresh forage that we cut with a scythe or reel mower. About that time, I was expanding my poultry flock, acquiring more egg customers and needing more housing. Dad suggested that I use the now-defunct rabbit runs as portable chicken pens. We took out the middle partition and the first portable chicken pen was born. That was about 1970.

An amazing thing happened when we put the layers in the pen and began moving them around the pasture. Their feed consumption dropped considerably, even though they had always had a yard in which to run. We began to see the importance of a daily fresh salad to stimulate consumption. Nobody likes a plate of stale food - not even chickens.

In the winter, we moved the chickens and the pens into the barn on mesh floors. Immediately, feed consumption increased by 30 percent. In early spring, we would move the birds back outside on the grass.

In order to sell more eggs, we began marketing them at a local farmer's market, called the "Curb Market." Begun during the depression as a way for farm families to earn cash, by the time I began selling there in 1970 as a 13-year-old it had dwindled to only two vendors: two elderly women. I joined 4-H in order to sell there, and that exempted us from all inspection requirements. We were under the auspices of the Extension Service, which had arranged the exemption with the state inspection department.

One lady sold baked goods (that's where I discovered my affinity for pound cake and potato salad with sweet pickles in it). The other lady came from a more diversified homestead and sold pork, baked goods and vegetables. At the time, we were milking a couple of Guernsey cows by hand. We sold eggs, beef, pork, rabbit, chicken, vegetables, butter, cottage cheese and buttermilk. The market opened at 6 a.m. on Saturdays and closed about 11 a.m.

As I culled old laying hens, I butchered them, cooked them in a large roaster pan, picked off the meat, packed it in quart freezer containers, and sold it as precooked, processed chicken at the Curb Market. An ice-cooled display case kept things cold. This meat went like hotcakes. I well remember at the time that the going price for spent hens was 19 cents, and mine were yielding nearly $2.00. Certainly I had processing time and money in them, but not that much.

As the flock increased, I needed a more efficient way to butcher. I ran an ad in the rural electric cooperative magazine for a chicken picker (I didn't know they were called pickers. I called it a plucker) and received a response. We hitched a trailer to our 1963 Chrysler Newport and drove across the state to buy an old table top type from a man on the Eastern Shore. With that picker, I could come home from school and butcher

21

thirty old hens in a couple of hours. My able assistant (slave) was my sister, Loretta, who was five years younger than I.

The Curb Market experience taught me marketing skills and business acumen. Dad taught me early the need to make a profit, so that I could replace my inventory of hens as they wore out. I developed a great appreciation for consumers, for selling a product that exceeds the customers' expectations, and for competitive pricing. That was in the days before health awareness hit the mainstream, when people were still eating butter and beef. Farm chemical usage hadn't even become a vogue topic yet. I also learned how much more money can be made by value adding to the product and taking it directly to the customer.

Dad viewed this as an excellent opportunity to build a direct market business free of encumbering inspection requirements, providing higher quality food than conventionally-produced food, but at a competitive price. When I went to college, we shut the booth down, I liquidated my 300 laying hens, and unknowingly, hammered the last nail in the curb market coffin.

Four years later, the "Curb Market" was gone. Apparently, the two elderly matrons had been holding on, inspired by the enthusiasm of a teenager. I often wonder what could have been had we kept the booth open. Our "Grandfather Clause" protection may have given us opportunities practically unheard of in today's regulation-happy environment.

Even as a child, I dreamed of farming full-time. Dad and Mom both worked off the farm to pay for it, and spent the early years doing research and development and conservation work. After graduating from college with a Bachelor's degree in English, I worked as a full-time reporter for our local daily newspaper for three years before coming to the farm full-time September 24, 1982. Teresa and I lived in an attic apartment in the farmhouse, drove a 15-year-old car, did not have a television, and committed ourselves to a dream.

People who want to farm must be committed enough to sacrifice for it. Young people who want to farm could even devote their time and energy to a farmer they know, working for free if necessary to show their character and commitment, to merit being entrusted with a farm. Often, however, teenagers are too busy buying that late model car, the newest home entertainment gadget, renting the latest video and visiting the latest trendy clothing store. There is neither time nor energy to pursue a vision.

Dad and Mom paid for the farm and gave us a running start, but we took the foundation and laid a profitable business enterprise on it, a business that did not become lucrative until several years down the road.

Opportunities still exist for those willing to go against conventional thought and commit themselves to a dream. I trust that this book will inspire you in new directions, with renewed commitment.

IN THE BEGINNING

Chapter 5

Getting Started

A pastured poultry enterprise is ideally suited to piecemeal starting. This spreads capital investment over time and reduces catastrophic failures.

The philosophy today to start a business is to find investors, borrow a pile of money, and jump into the enterprise. Few people start in the back closet with a few extra dollars and let the business grow slowly into a full-time enterprise. But pastured poultry can easily be done that way.

I suggest that you start with 50 or 100 broilers. We started with 400 and that was too many. The second year we backed off to 300. The third year we jumped to 800, but that was the third year.

Don't worry about selling any from the initial production. Plan to eat them all yourself. If they are nice, give away a few. But that will just be cream. You would be amazed at the things that can go wrong, so the most reasonable thing to do is assume that you'll have problems and be pleasantly surprised if in fact you get a good portion of those initial birds up to slaughter.

You will get your chicks from a hatchery, probably air freight through the mail unless you live within driving distance of a hatchery like we do. Don't push the season. Remember that cold is one of the hardest things on baby chicks. Start your first batch after they can be assured postal handling at 50 degrees in nonheated trucks. If you have to wait until late May or June in order to do that, do it. Believe me, the wait will be well worth it. Once you are an expert, then go ahead and push them all you want. But I advise against it initially. Your initial experience should be rewarding and pleasant, and the best way to ensure that is to minimize the potential risks.

When the chicks arrive, you need to get them warm (90° F) as soon as possible. If they are cold, they will not eat. Their first desire is warmth - that supersedes everything else. For 100 chicks, an area 10 feet square is adequate. A couple of cardboard boxes in your house, with a light bulb hanging down, is fine. Use a thermometer to check the temperature at chick level. I know one lady who put the box on top of the refrigerator where it was warm. You can use a woodstove as well, or a furnace in the basement.

You will need a waterer and feeder, but initially just some newspaper with feed and grit sprinkled on it is fine. You want unlimited access to both feed and water. Use dry sawdust or shavings for bedding. If you put the chicks in the house, remember to control the cat and dog, for obvious reasons.

One pen is adequate for this number of birds, and I suggest that you construct it prior to receiving the chicks. The temptation is to go ahead and get the chicks. The inevitable delays that accompany research and development will occur and before you know it, the chicks will be crowded and you won't be ready to move them outside.

The feed can be a problem if you are a purist and do not want to use commercially manufactured feed. Most mills will not custom mix less than 1000 pounds of feed. I suggest that you just use the regular chick feed. It will have antibiotics in it, of course, but you have to start somewhere. Get the least medicated possible. Realize that this is a temporary thing. Next time, when you step up to a couple hundred birds, you can get your ration custom blended.

The only other solution is to get your feed from someone who is already raising poultry with our ration. Several folks in our area come and get feed out of our tanks for their small flocks. We certainly do not want to get into the feed business, but we are glad to help a person who is trying to produce his own food.

With the chicks up and going and out in the pen, you are pretty well set until processing. Obviously, this too needs to be planned before the chicks arrive.

You can process the birds by hand or purchase some very reasonable small scale processing equipment. If you do them by hand, invest in a submersible thermometer so that you can be sure your water is at 145° F. You can use a pot over a wood fire, or rig up a gas burner, or plumb an electric element in a metal barrel. I can't begin to suggest all the ways to rig up a scalder. Only you know what resources you have.

You can hang the birds on a piece of string for picking. That would be the simplest way, if you are picking them by hand. Water and a clean table surface are necessary for the eviscerating.

This is as simple a startup plan as I can imagine.

Simple home-sized processing equipment, though, is available at a reasonable price. I consider the minimal processing equipment to be a 5-gallon neoprene thermostatically controlled scalder and a tabletop picker. These can be had for about $300 and will allow a couple of people to process 10-20 birds per hour. Unless you are an expert eviscerator, you won't be able to go faster than that anyway.

People I know who have invested this much in the minimal equipment have no problem keeping it busy renting it out to other folks in the neighborhood who want to process their own poultry. Even if you decide chickens aren't for you, it is not a wasted investment. I suggest that you seriously consider investing this much, even for 50 or 100 birds, just because it will make the initial processing experience much more enjoyable. First impressions are important.

One of the beauties of poultry is that getting started is as simple as it is. It's nothing like starting in other livestock enterprises, which require substantial acreage, fences and facilities. The initial investment in a couple of cows or sheep is far more than getting started with a hundred broilers and some mechanical processing equipment.

One of the nice things about the pasture production model is that you need not build a "chicken house." This means that a "house and lot" person can run the birds in his back yard in the portable pen. It exempts you from building ordinances because the pen is a machine. Remember that one pen, moved daily during the 36-40 days the birds are outside, will only cover 4000-5000 square feet. Many people have that much in their yard. The point is that you need not own a farm, rent land, or do many of the other things that typical farming endeavors require.

If local animal ordinances are a problem, remember that bureaucrats generally spend all their time handling complaints. These chickens do not make any noise. What makes noise is mature roosters crowing. Chances are if you run the pen in your back yard, nobody will even know that you have a pen back there. If you have a complainer, offer to take him a good chicken - make him an ally. I guarantee you'll have the prettiest, greenest lawn around. Perhaps you can start a new fad in your neighborhood. I think few things would be as gratifying as seeing neighborhoods all over the country run a chicken pen in their yard to mow it, fertilize it, and produce their own chickens. How exciting.

Anyway, the biggest hurdle to doing some-thing is to start. We all know that most jobs are not nearly as bad as they seem. The hard part is starting. Many people ask me how to get started farming. My response is, "Well, what are you doing right now?" Chances are, all they are doing is stewing about doing something. Channel your frustrations into something positive. If you wait for a farm to be dropped in your lap before you start a garden, or raise some chickens or rabbits, either at your place or someone else's, you will never do anything in farming.

I guarantee you someone in your circle of acquaintances could use a hand making some hay, working some cattle, chopping thistles, or building a fence. You need to make a genuine commitment to doing something, anything, with or without remuneration. The *Bible* calls this casting your bread upon the waters. We have too many people, especially young people in agriculture classes, who are more interested in the retirement plan than in getting involved in agriculture to help someone meet a need. We have too much selfishness and not enough service. If they would spend their time helping a farmer instead of driving their car, watching videos and complaining about the lack of opportunities, they would have opportunities galore.

Believe me, few people get a farm dumped into their lap. If a dream is worthwhile it is worth some sacrifice, and that starts with little things before progressing to bigger things. And it starts with you, not the other guy. Too many people want to get the bigger thing before they've proven faithfulness in the little things.

The main hurdle to realizing the full benefits of this pastured poultry opportunity is not the work, the marketing, the processing or any other facet of the enterprise. It is getting, producing, and processing the first chicken. In short, the biggest hurdle is inertia. Once you conquer that, you're off and running.

The point of this discussion is to encourage you to start small, to start with something, and to realize how little capital and time is required to raise a few chickens. It really is not that complicated.

Go for it.

Chapter 6

Choosing A Breed

As with all species, different breeds contribute certain assets and liabilities. A strong point in one breed will be offset by a weak point in another area. Chickens are no different.

The old standard American varieties like Rhode Island Reds, White Rocks, and New Hampshire Reds are called dual purpose breeds. They lay fairly well and yet are deep bodied and yield a fairly meaty carcass. But modern genetic selection, especially with the onset of the vertically integrated broiler and egg industries, has further specialized breeds and even strains within breeds so that virtually no commercial flocks now consist of the old dual purpose breeds.

Hatcheries have even hyped up the old breeds, especially the Rhode Island Red, perhaps the most popular, into a strain called the Production Red, which is a better layer but a smaller bodied bird. The White Leghorn is undoubtedly the queen of the commercial egg industry, weighing hardly 3 pounds as an adult and laying more than 300 eggs in a year.

On the meat end, the reigning king is the

Cornish Cross. When this hybrid was being developed, it was a cross of the naturally double-breasted Cornwall chicken from England and the tall standing White Rock. The White Rock gave skeletal height and frame to the short, squatty Cornish chicken. Virtually all the broilers in America are Cornish Cross.

This hybrid, constantly being genetically upgraded by the commercial industry to gain faster and convert grain to meat more efficiently, is not without problems. Bred up to perform at a totally unnatural gain rate, this bird is prone to all sorts of diseases and structural deformities, and is in all ways an extremely fragile bird.

I call these race car chickens. When everything is adjusted just right, they are awesome. But it doesn't take much to throw things out of adjustment and throw the car into a pit stop. When these birds are doing well, their performance is unparalleled.

The old nonhybrid varieties are hardy and mortality runs close to zero. The reason they are not used is because they grow much more slowly and their breasts are not broad, but rather sharp and thin. As a result, the meat is not nearly as palatable because to get a 4-pound carcass these birds must be about 12 weeks old as opposed to 7 or 8 weeks on a Cornish Cross.

The older the bird, the tougher the meat. When it takes longer to produce a marketable carcass, palatability goes down. This difference in the meat is a geometric progression, meaning that each succeeding week beyond a certain age has a greater effect on meat quality. I hate to say quality, because the older the meat the more complex the amino acids and therefore the more nutritious it is. But in terms of "melt-in-your-mouth" consistency, the meat deteriorates rapidly beyond the ninth week of production. The industry continues striving for earlier and earlier processing weights and ages, which militates against nutrition. Of course, if everyone cooked their meat in a crockpot, this whole argument would be unnecessary. That is the way to make even tough meat tender - long, slow cooking. But that is not the way the average American cooks meat. Market acceptance, therefore, puts us on the teeter-totter, balancing young and tender on one end and the best nutrition on the other.

Perhaps one of the biggest reasons to use the Cornish Cross is marketability. The American consumer has grown accustomed to the double breast and the overly meaty carcass; the appearance of alternative, razor-breasted carcasses is hard to accept. Certainly some consumers will buy the tougher, narrow breasted types, but the market niche is too small for the small scale producer to make a living at raising poultry.

Of course, consumers can be educated to the merits of smaller, narrow-breasted birds if you mount an aggressive teaching campaign. This would include raising a few of the alternative birds on the side, giving away samples, and perhaps offering recipes and cooking instructions that would encourage a successful culinary and dining experience. A market for the nonhybride can certainly be cultivated, but it will be much more difficult to develop than the Cornish Cross market.

Consumer acceptance is extremely crucial in marketing. The average consumer can only be pushed so far, and for the unconventional producer, enough new ideas are being thrown at the

33

consumer without offering the notion that chickens ought to look more like quail than butterballs.

We've opted for the Cornish Cross and have chosen to fight the deficiencies nutritionally and with good management. For feed conversion and quick cash turnaround it is unprecedented. Consumer acceptance is tremendous and processing is much easier because the hair has been bred off these birds and they do not require singeing after feather picking. The white pin feathers are not as easily seen as are colored pin feathers. They are also the most available of all the breeds, since virtually every hatchery has them. Genetically they are more consistent than any other variety, from hatchery to hatchery.

Purchase your chickens from a hatchery that seems honest and one that gives the most consistently hardy chicks. Hatcheries are scattered all over the country. I have used several over the years and found more differences in postal connections than in the quality of the birds. I think most hatcheries deliver an acceptable product. I suggest you try several and patronize the one that has the best postal connections and the highest quality chicks.

Chapter 7

The Brooder

Chicks hatched without the benefit of mother hens require careful attention because they are quite fragile. They must feel as secure and cared for as they would if their mother were there to encircle them with her wings for warmth and protection.

The floor space needs to be roughly 25 square feet for 100 birds up to 4 weeks of age. They can certainly be confined tighter for the first week.

The area needs to be predator- (including rat-) proof and free of drafts. Nothing can decimate a flock of chicks faster than predators or chill coming from drafts. The final primary requirement is dryness. Chickens can't take dampness.

The brooder should allow as much natural sunshine as possible since this stimulates the pituitary gland far more than artificial light. Full spectrum light is a desirable luxury. Be sure to put skylights in the roof to encourage light and eliminate light bulbs. Translucent fiberglass panels work well for this.

Day old chicks need to have access to 90°

heat. The entire floor area need not be that temperature, but there must be enough heated area to allow all the chicks access at one time. They will run out from under the heated portions to eat and drink, and then run under to sleep.

The actual heat source can be as elaborate as thermostatically controlled units powered by electricity or natural gas, to a wood stove to a simple heat lamp or light bulb. Generally, insulated hoods are used in conjunction with the heat source to make it more efficient. Thermometers need to be handy to monitor floor temperature. Feed sack skirts around the hover bottom help hold in heat.

The floor should be covered with dry flakes of some carbonaceous material. I prefer wood shavings from fine woodworking shops because they are always extremely dry and quite small. Many materials do not work. Hay and straw tend to mat quickly, reducing manure penetration and forming a damp, filthy layer on top. Wood chips are too big for the chickens to scratch around, thereby producing the same effect as hay and straw. Leaves mat down and are also hard to get dry enough to absorb the moisture. The wood shavings are fine enough to afford the chicks plenty of scratching, fluffing and beak cleaning, which they do by swiping the beak on the litter, first one side, then the other.

A wire mesh floor should not be used because it does not let the birds scratch, fluff or clean their beaks. It is devoid of living beneficial microorganisms that contribute much to the health of the tiny birds.

We find a direct relationship between mortality and quantity/quality of the bedding. When the brooder is cleaned out between batches and new bedding put in, the mortality is higher than if it is not cleaned out but only aerated, with new material added to the top. The carbon-nitrogen ratio should be about 30:1. The quickest way to determine if the ratio is about right is through smell. If you can smell ammonia, there is not enough carbon to soak it up. Raw poultry manure is about 7:1; cow manure 18:1; wood shavings 500:1.

Carbon bonds to the soluble nutrients and holds them in chemical suspension. Any time you can smell ammonia in a livestock facility of any kind, be assured that more carbon must be added. When the animals breathe in that air, it not only harms tender respiratory mucous membranes, but pulls vitamins A and D out of the body, depositing them in the liver at toxic doses. Damaged livers are common today in the confinement poultry industry. Fecal and nitrate contamination in feeders, waterers and air particulates are escalating problems with intensive confinement housing. To show how out of balance these conventional models are, the carbon-nitrogen ratios generally run about 12:1 instead of 30:1. That is why the bedding is toxic, won't compost (even though it heats) and pollutes both air and water.

If the carbon-nitrogen ratio is correct, the bedding will actually compost if air is injected into it. Between batches, therefore, I wet down the litter with water and then take a hoe or shovel and stir the bedding. In a big operation this could be done with a rototiller. Remember that it will not work if the ratio is out of whack. Then I just add a couple inches of fresh wood shavings on top and put in the chicks. We literally take out the 2-3 week old chicks one hour and put in new ones the next. No cleanout, sanitizers, germicides and the like are ever used in our facilities.

The deeper the bedding, the better. The composting action is warm, which reduces the need for heat. It's an even heat, too, and extremely comfortable for the chicks. We've actually raised several batches with virtually no brooder heat. We built a new brooder facility to accommodate an 18-inch bedding depth.

The deep bedding grows natural antibiotics through the molds and fungi, grows a tremendous number of little bugs that offer protein for the birds, and gives off warm, moist heat very much like a mother hen's. The composting reduces the overall mass of the bedding, so that when cleanout is necessary the total material is a fraction of what was originally put in. Common shrinkage is 30-50 percent. To be sure everything is correct, scoop up a handful of the bedding from deep down and make sure it smells like sweet, fertile soil or moldy leaves. It should not smell like raw manure.

The brooder temperature, after 48 hours, can be reduced a couple of degrees per day for 3 or 4 days, then 3 or 4 degrees per day in an accelerating curve (see Figure 7-1). By the time the chicks are three weeks old they can handle freezing temperatures if the air is still and dry. They are actually much stronger than most commercial producers realize. It is important to decrease their brooder temperatures early so that they will develop the hardiness necessary to survive outside on pasture. The procedure parallels hardening off greenhouse plants prior to transplanting.

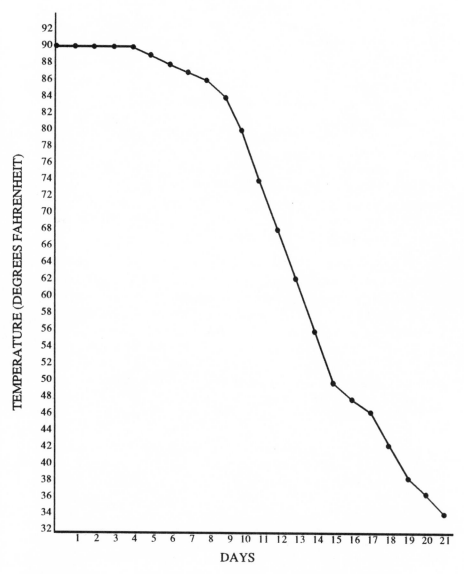

Figure 7-1. *Temperature Requirements for Chicks*

The birds need enough linear feeder and waterer space to accommodate 35-50 percent of the flock at once. Just like babies of any other species, chicks spend the bulk of their time sleeping. It is unnecessary to have enough feeding and watering room to accommodate them all at once.

It is imperative, however, to give them plenty of feeder space. When the birds line up at the feeder, just after you've put in fresh feed, count the chicks on the side and compute how many birds can get around the feeders. If it is not at least 35 percent of the total number, add more feeders. Remember that every few days this amount will change, so you need to be sensitive to the growth of the chicks and the fewer number that can get around the feeders.

The reason this is so critical is that one of the easiest ways to foster leg problems is inadequate feeder space. When a chick can't get feed easily, the stress will show up in its legs. In addition, make sure the chicks never run out of feed. That is one of the worse things you can do. It is better to waste some feed than to let them run out of feed. Remember that these are high metabolizing animals. A little chick is just like a human baby in that it feeds several times throughout the day. No more than you would routinely ask a human baby to miss feedings, you should not ask chicks to miss feedings. Give them plenty of feeder space and make sure they have all they want all the time.

The brooder facility needs to be tight when a cool breeze blows, but also well ventilated when the weather is hot and muggy. Most facilities use a hinged window or translucent curtains, with hardware cloth behind, permanently affixed to the structure. The brooder house need not be high

40

enough to walk in. We've built some 8 ft. x 10 ft. brooder huts with a peaked roof made of two plywood panels hinged at the peak (Photo 7-1). The bottom is an 18-inch high box made of pressure treated 1-inch x 6-inch lumber. A dirt floor is fine as long as it is adequately covered with dry shavings.

If you are running more than 200 or 300 chicks, the brooder should be partitioned so that the birds can be sectioned off in groups of no more than 300. Performance in poultry always drops when the number of individuals in any one group exceeds 300.

When we run 1200 chicks, therefore, we section them off into four groups, each with its own hover, feeders, waterers and grit material.

Photo 7-1. *A simple brooder hut: just a box on the ground. Also used as the hospital pen.*

41

If something frightens the birds, the pile-up pressure will not be as great as when the numbers are small. If a thousand chicks flee to one end of the brooder, be assured a bunch of birds on the bottom will suffocate in a minute. The competitive pressure on the birds is simply not as high when the group is smaller.

Commercial producers use pre-made folding pieces of heavy cardboard and then remove them in the second and third weeks as the birds get acclimated. We have found that the partitions are just as critical at three weeks as they are in the first week. One way to use the partitions is to confine the birds more tightly in the first week and then gradually move the partition back to give them more room as they grow. But the main partitions between groups of 300 should not be removed.

I built partitions out of plywood, hinging the pieces so that they stand up like an accordion (Photo 7-2). The partitions need not be high - 18 inches is plenty. By the time the birds can jump up on top of it they are ready to go out to pasture anyway. You want the partitions light and short to allow portability and access. The main thing is to keep these birds in small groups from day one all the way through to processing.

A lady told me once she used oval cattle watering troughs as brooders. Ideally square corners should be transected with short boards to reduce corners where chicks can pile up and suffocate. The structure may be as simple or as elaborate as imaginable, but it must be capable of warmth, dryness, ventilation and be closed to drafts and predators.

Photo 7-2. *Loading chicks out of the brooder to go to the field.*

Photo 7-3. *Main brooder, 12 ft. x 20 ft., on pole skids so we can move it if necessary. Solid wooden floor.*

Chapter 8

Starting the Chicks

When the chicks arrive in their shipping boxes, they can be cold, thirsty and hungry. If they are warm, they will want to eat and drink immediately, and probably explore their brooder facility. If they are cold, they will want to get warm before doing anything.

Baby chicks can generally go 72 hours without food or water. If they are cold, be patient and let them get warm. Within 24 hours after arriving, a hundred chicks should drink about a quart of water. If they drink, they will eat. To encourage eating, spread out some newspapers and sprinkle feed on top so that they can have quick, easy access early on. After a day or two of that, take away the newspapers and just allow access to the feed troughs.

It is important to have different size feed troughs. Chicks need small ones and larger birds should have a bigger one. Actually, the height of the trough lip is the critical factor.

You don't want to make the chicks stretch to get their feed. In fact, most trough type feeders should be nested into the bedding a half inch or so to make sure that the lip does not exceed beak height. You want the birds to be able to reach down and into the feeder without stretching too much. That constant stretching can cause leg problems. Better to let them waste a little bit of feed than to encourage crippling.

As the chicks scratch away the bedding from the feeder, it will rise in relation to the floor and therefore the chicks. Just move the feeder over a couple of inches so it is back on the scratched, settled bedding. You want the lip of the feed trough to hit the birds at about breast height. If it is too low, they will scoop out feed and waste it. If it is too high, they have to stretch too much for it and will cripple.

I prefer trough feeders to round feeders. The trough provides more linear space in relation to the amount of feed it holds. Round troughs are prone to bridging. They work fine for pelleted feed, which has a lot of air space between particles, but mash bridges inside and won't fall into the pan.

Waterers should be elevated as the chicks grow, not only for cleanliness, but also for drinking efficiency. The waterer lip should come to about the bottom of the chicks' beaks. The waterer lip should be higher than the feed trough lip.

I believe it is extremely important to get silica and grit into the gizzard as soon as possible. I do not recommend commercially manufactured grit but rather creek sand and aggregate, finding it much more diverse in mineral content. It contains particles of greater dissimilarity in size, and also contains bits of vegetable matter like roots, bugs and other things only the chicks can see. Generally I sprinkle some of this grit on the newspaper and then sprinkle the feed on top to ensure that all the chicks, right away, receive a healthy dose of grinding stones for their gizzards. If the gizzard has plenty of grinding material it will work more efficiently.

If possible, hay chaff, in a separate pan, can be fed to the chicks because the seeds of perennials and weeds are more nutritious than are the seeds of annuals (feed grains). We carry a wash tub on the hay wagon when baling hay and let the chaff from the baler chute fall into it. In the course of a season we can accumulate barrels of chaff. Chicks eat many pounds of it and learn quickly to scratch and search for tidbits of food. What they don't eat simply augments the bedding.

These scratching and forage eating skills not only stimulate pasturing ability, but also afford exercise opportunities for strong structural development. With every passing day, these chicks become more lethargic. Their meat growth outpaces the development of their skeleton and internal organs, which decreases their physical activity.

Any vegetable matter that is in season is a good supplement. Things like lawn clippings, ragweed seeds, pigweed seeds, and dandelion blossoms all give the chicks fresh nutritional supplementation and stimulate their desire for forages. These must not be given in excess, however, because they are low in energy and can be detrimental to the birds' development if consumed to the exclusion of grain.

Remember, green material is high in vitamins and minerals, but low in energy. Furthermore, we are dealing with a bird that the industry has bred to require a certain level of "hot" feed in order to perform. In other words, if we fed these birds like the settlers fed their chickens at Williamsburg 300 years ago, many would probably die.

I think it is unfortunate that we have created a chicken that is so far removed from a normal chicken's ability to forage and fend for itself in the barnyard. There ought to be some middle ground. But the fact is that we are dealing with a chicken that was bred to eat high-calorie, low vitamin and mineral feed without fresh air and sunlight, on antibiotics and hormones. We can do many things to make this bird worth eating, but there is a limit to what we can do. This has been the downfall of many organic producers who do not want to feed animal byproducts like fish meal or meat and bone meal because of the residues left from slaughtering the parent animal material. I think this is throwing out the baby with the bathwater. Remember that birds naturally eat large amounts of animal protein - carcasses, bugs, worms and insects.

If that material is unacceptable, then I suggest you not try to raise Cornish Cross birds. They will be sickly, prone to all sorts of skeletal problems like crooked feet, and many will simply die. We must feed the bird within its genetic parameters. And a diet of nothing but grain will simply not do the job with these hyped-up race car birds. They need a higher octane fuel.

You can actually feed so many lawn clippings, for example, that the birds will consume too much of it and get sick simply because they are replacing too much of their high octane feed with low octane - albeit high in vitamins and minerals - green material. These supplements need to be fed as supplements, much as you would take supplement vitamin capsules to augment your meal. But you certainly would not eliminate your meal. Give them what they will eat in 10 or 20 minutes, then back off and give them some more tomorrow. The law of the little bit applies here.

At this point it is necessary to point out something obvious: pastured poultry absolutely must NOT be debeaked. This procedure reduces cannibalism in factory confinement houses. If raised properly, chickens will not peck each other. Cannibalism is such a nonproblem for us that we have actually had birds wounded from predators or injury, heal up and not be picked on by the rest of the birds. Cannibalism is caused by a nutritional inadequacy primarily, and boredom secondarily. Stress of some kind is the overriding cause.

Debeaking renders birds nearly useless for eating anything but prepared mash. In order to clip off and ingest a 6-inch blade of grass, a chicken needs that full, graceful beak. Nothing less will do.

Visual observation is extremely important for raising healthy chicks. If the birds spread to the edges of the brooder house, it is too hot. If they bunch up under the heat sources and pile up, it is too cold. They should routinely stretch themselves, extending a leg back and at the same time extending a wing up as high as it will go. They should walk erect and tall, not hunched over with a stilted gait.

When you walk outside the brooder house, you should hear the happy chirping of the chicks. If they are quiet, something is wrong. They always sing if they are happy.

Keep a liberal amount of fresh, dry wood shavings handy to immediately place on damp spots, especially around waterers and under brooder lamps where the chicks lounge. After several days, the entire floor should be recovered with a half inch of fresh shavings, or at least stirred with a hoe or rake to turn up fresh litter if it is clean underneath. Since these birds spend much time sleeping on the floor, their joints are extremely susceptible to problems caused by floor dampness.

Bedding need not be changed between batches of chicks in the brooder facility. Much research suggests, and my own experiences attest to the fact that it is better to let the litter build up than to clean it all out. The facility need not be sterilized in any way. Deep litter of 4 to 8 inches, properly aerated, produces natural vitamins and antibodies through molds and fungi. The litter should not smell like ammonia. If a chicken house ever smells of ammonia, the litter is either too wet or does not have enough carbon in it. The carbon bonds molecularly to the highly soluble ammonia, locking it up in a chemical sponge suspension. That is the beauty of deep bedding.

Normally, about 1 or 2 percent of the chicks will die in the first couple of days. These are generally birds that were sick from the hatchery or were runts and just never got going. If you lose 4 or 5 percent in the first week, something is wrong.

This level of care for baby chicks may sound too high for some people, but it is necessary in order to not only prepare the birds for efficient pasturing, but also in order to raise healthy birds without vaccinations, toxic housing, sterilizers and antibiotics. Drugs, sterilizers and debeaking machines merely mask maladies caused by inappropriate production models. To raise a nutritious chicken, it is not enough to eliminate the negatives. We must institute positive procedures to stimulate the chickens' own abilities to thrive.

As you become skillful at raising the chicks, you will be surprised what a cursory glance can tell you. In 10 seconds I can tell if the birds are happy, if they have plenty of accessible feed and water, if they are under stress. It's like any acquired skill - there's no substitute for experience.

Chapter 9

Ration

Perhaps nothing is more crucial to the performance of the broiler than what the bird eats. One beauty of pasturing is that the prepared ration represents only about 80 percent of the bird's diet, as opposed to 100 percent in a typical confinement house.

As a result, we've found that we can fluctuate, or "play around with" the ration quite a bit before we see significant performance changes. At the outset, then, realize that our ration is good for us in our part of the country at our local prices. It may or may not be the best for someone in another areas. Not being an expert in feedstuffs, I will not offer every alternative, hoping that you will try things small and be the final judge.

Our ration began as a fairly conventional but unmedicated broiler mix without hormones, many years ago. Over the years, we gradually eliminated artificial vitamins and acidulated minerals, substituting more natural supplements. A typical ration may contain 30 or more ingredients, and normally, the commercial broiler industry changes the ration every week of the bird's life. Basically this change represents a gradual shift from protein to starch.

The baby chicks require the highest level of protein. As the bird matures, the protein requirement diminishes and the need for starch increases as a percentage of the diet. Starch, or energy, is relatively inexpensive compared to protein. The incentive, therefore, is to get by on as little protein as necessary and pump the energy to these birds. The protein, however, is what builds feather and muscle (meat). True health requires more than just a life-sustaining minimum.

Here is our most current ration as of 1998:

INGREDIENT	POUNDS/PERCENT
CORN	52
ROASTED SOYBEANS	29
CRIMPED OATS	11
FEED GRADE LIMESTONE	1
FERTRELL NUTRI-BALANCER	3
FISH MEAL (SEA-LAC)	3.5
KELP MEAL	0.5
FASTRACK PROBIOTIC	0.1
TOTAL	**100.1**

I'll go through each of the ingredients and explain why it is in the ration. By far the biggest ingredient is **corn.** As noted earlier, this is energy, and these birds do need energy. They run, scratch, grow and flap their wings. Pure caloric requirements are incredibly high on these high performance broilers.

In many areas of the country, milo or grain sorghum can be substituted for the corn. The conventional poultry industry routinely uses train car loads of waste oils and grease from food processing factories as a cheap energy component. Much of this grease comes from poultry processing. While I do not recommend this type of oil or grease inasmuch it is from poultry stock, I see nothing wrong with using waste kitchen oils or grease from vegetables or other meats as a tonic for broilers.

Any high fat supplement, used only as a tonic, can be beneficial for these high octane birds. One of my favorites is clabber milk. If you can get extra milk, let it sour and then feed the birds the congealed portion. These curds provide a wonderful tonic to growing birds before their sixth week. After that time, curds add too much fat to the carcass.

We crack half the corn and grind the other half in order to give some texture to the feed and make it easier to swallow. Imagine trying to swallow straight flour. Rations too finely ground cause blockages in the esophagus. Although the big pieces may be a little large for chicks, that's fine because it will automatically increase the protein percentage for the first week as they leave the bigger corn pieces. Then they will begin eating the leftover corn pieces in the second and third week, bringing the protein down a bit and eliminating the wastage.

The second component is **roasted soybeans,** the

protein. We have ready access to roasting, which preserves the fat and oils. These oils are extremely high in vitamins. At 20 percent fat, roasted beans taste delicious both to humans and chickens. The primary reason for roasting is to get some additional vitamins and oils without sacrificing protein.

Certainly you may use soybean meal or even cottonseed meal, if necessary. The key here is vegetable protein rather than animal protein. One of the greatest benefits of pastured poultry is no cannibalism, whether it consists of the birds eating on each other or through a ration containing poultry parts -- cooked, processed or otherwise.

All sources of animal proteins like bone meal, blood meal and meat meal have fallen into disrepute due to their association with mad cow disease and other diet-enhanced maladies. Although the conventional mindset views protein as protein, regardless of source, we in the clean food movement do consider source as not only relevant, but actually important. The toxicity, both in terms of animal diseases and pharmaceutical residues present in meat by-products from the factory farming sector makes these components off limits to pastured poultry producers.

The third component is **crimped oats.** Whole oats are fine, but crimping makes them a little more digestible. Although oats are not considered nutritionally prime, they offer other properties just as important. To quote from Morrison's *Feed and Feeding,* "because of the hulls, the use of oats in poultry rations tends to prevent feather picking and cannibalism. Also, the hulls seem to provide a factor that improves the growth and feather development of chicks and helps prevent mortality."

The oats and hulls also reduce the dustiness of the feed, making it easier to ingest and handle. When feeding or storing in a hopper-type container, the oats add enough pieces to make the mix flow rather than sticking and bridging. A finely ground ration clumps together and will not drop as easily in a hopper. Larger particles add more air to the mass, encouraging flow-ability.

Certainly wheat, barley or rye can be substituted for the oats, but we have found that neither rye nor barley are as palatable. Chickens certainly like wheat, but it does not have the hull benefits of the oats. Remember that a bird's natural diet is heavy on seeds, many of which have a heavy hull. This fiber is important in proper digestion.

Fish meal is an extremely expensive component but it complements the grains with high vitamin and mineral contents. It will often run about 60 percent protein, and the high vitamins A and D content, as well as the 18 percent mineral/matter ratio, makes it an excellent all around supplement for poultry rations.

The trademarked product Sea-Lac, which we prefer, is manufactured using low heat. This preserves a higher portion of the goodies. Fish meal is not considered an animal protein since it comes from the ocean and is not as subject to the concentration problems of factory farming. Heavy metal toxicity, especially mercury, has been a concern in aquatic-based feed supplements, but these have generally been highly localized and close to shore.

Whatever small risk we take in using fish meal is more than offset by the high amounts of chlorophyll the birds ingest. Chlorophyll is the

number one natural cleanser and detoxicant, perhaps next to humus.

We have many environmentally sensitive customers, and so far we have never had one react to
our chicken. Some have violent reactions to those
produced conventionally. The point is that we can
ameliorate any possible negatives from the fish
meal with the proper counterattack and still get
the benefits this supplement has to offer. Obviously, we'd love to find a substitute, but so far
have been unable to find one. Instead of throwing
out the baby with the bath water, we try to neu-
tralize the negatives in order to capitalize on the
positives.

Fertrell's Nutri-Balancer is a mineral poultry supplement based on the thesis that the modern-
day love affair with synthetic proteins and ani-
mal-based proteins are the result of everything
being demineralized, from grains to rations. According to this reasoning, perhaps best articulated by Jerry Brunetti, internationally-known consultant and lecturer on the subject, animals are
unable to access proteins in feedstuffs for primarily two reasons.

First, chemical fertilization, hybridization
and genetic manipulation all reduce the minerals
in grains. This creates a mineral shortfall coming
into the trough. Second, the minerals facilitate
protein assimilation. If the minerals are not in
the feed's raw ingredients, they must be supplied,
in proper form, as a supplement. Then the gut can
metabolize the protein in the feedstuffs.

The importance of minerals in the food chain
is fast becoming the darling of the alternative
agriculture community, somewhat like organic mat-

ter was twenty years ago. Of course, the organic community never agreed with the simplistic NPK philosophy, but the new interest in raw minerals goes beyond merely defending trace minerals. Even raw salt products now line health food store shelves.

The backbone of Nutri-Balancer is dicalcium phosphate and soft rock phosphate. In addition, it contains high levels of vitamins A, D3 and E, as well as trace minerals. The Fertrell company, headquartered in Bainbridge, PA, has been a premier formulator and supplier of natural soil amendments for decades. The company is now entering the livestock feed supplement world with mineralized concentrates as a natural next step to what it has done agronomically.

Although this supplement is quite expensive, it has eliminated crippling and leg problems. In fact, we do not even have a hospital pen anymore. The meat tastes sweeter and the birds gain beautifully. One added benefit is that it is clearly producing a more balanced manure. Since eliminating animal proteins and shifting to Nutri-Balancer, we've seen a marked improvement in the composition and quantity of our pastures following the chickens. It is almost like getting paid $2,000 an acre to remineralize our farm.

Feed grade limestone provides additional calcium. Aragonite is probably the premier product to use for this since the calcium is more highly available, but it is not as common in the feed mill industry.

Because we believe in the supremacy of Icelandic geothermally dried **kelp meal**, we still use a tiny portion of it in our feed, even though the Nutri-Balancer contains kelp. Kelp contains more than 53 trace minerals, all in perfect sea brine balance. Authorities on organic foods and alternative health care consistently agree that it is not the amount of minerals in the body, but rather the ratio of one to the other that determines efficacy. No laboratory can succeed in a balanced mineral approach for every farm.

But nature, through sea brine, has preserved that perfect balance through seaweed. Sea brine mirrors the same mineral ratios as are found in healthy human blood. Ideally, this seaweed should be grown in the most stressful conditions possible (cold) and dried geothermally rather than with natural gas. The slow, low heat drying preserves more of the goodies than does the high heat of natural gas. Kelp also contains a multitude of vitamins and natural growth hormones.

Finally, a tiny amount of **probiotic** is helpful in fighting stress like heat, cold, and adult heart attacks. We use FASTRACK from Conklin Company, Inc. This is one ingredient that varies greatly from source to source. Some probiotics on the market are nothing more than hyped up trace minerals. The higher the percentage of colony forming microbial units per ounce, the better the product. The product we use contains no trace minerals, but is pure critters with enough carrier to keep everything alive. We use roughly two pounds per ton of feed, which is half the recommended feeding rate, because we've found the Nutri-Balancer puts the birds' health close to where it needs to be anyway.

Basically, probiotics contain *Lactobacillus acidophilus* yeast cultures. To quote a Conklin pamphlet, it is a "naturally occurring live microorganism that promotes proper digestion and increases the absorption of protein, starch and fat." Probiotics are not drugs, produce no resistance factor and leave no residues. They increase the effectiveness of the animal's own immune system.

We use the same ration from day one to the day of processing, partly because our production model allows greater flexibility before performance changes, but partly because it is simply easier logistically to deal with one ration. We find that the dollars saved by reducing protein on the final stretch are not worth the hassles of sick birds and keeping different batches of feed separate in the field.

Some technical advice may be helpful about grinding and mixing the feed. When you are mixing several tons of feed, the micro-supplements are hard to mix in evenly. In such cases a good feed mill will make a pre-mix of all the tiny ingredients, then add that to the big batch later.

For example, in our ration, the mill puts in kelp, feed grade limestone, Nutri-Balancer and probiotic, mixing that thoroughly before they add the first pound of feed. Then they add the other ingredients and mix it all together. The probiotic is too small to mix adequately if you just add the tiny amount to the rest of the feed. If your mill is not pre-mixing the small ingredients, they will be in little pockets, or globules, and not disseminated thoroughly in the ration.

Generally speaking, keeping time should be no more than three weeks once the feed is ground and mixed. As soon as a kernel of grain is opened, the nutritional value begins to deteriorate. It is better to get two batches of half a ton each than one batch of a ton that will be enough for the entire eight weeks. Sometimes you can save a little time or money by getting a larger amount of feed, but if you can't use it all up in four weeks, you're better off getting two smaller batches.

As with all my recommendations in this book, some are principial and immutable; others are subject to local conditions. For example, I'm not concerned whether you build the pen of PVC or wood, but it can't be too heavy to move every day. Likewise, I am a diehard fan of these high mineral supplements, and welcome all comers as long as the birds stay healthy and gain well. The grain portions can be adjusted for local availability, but the total protein should not fluctuate more than 2 percent.

A friend in Texas uses triticale instead of corn and oats. He's getting good results and his feed is half the cost of ours. Often, the best thing to do when getting started is simply to use a non-medicated bagged feed from the local mill. As you grow, you will want to get a custom mix. But do not be paranoid about the ration; the production model is the key to the bird. The thing to remember is to experiment. Do not be afraid of trying new items in the ration to see what works best for you. And don't just listen to the feed mill salesman. Do your own research and your own thinking. Most important, observe the birds and see how they perform as you fine tune the ration.

"Rational" Notes

_____

_____

_____

_____

_____

_____

_____

_____

_____

_____

_____

_____

_____

_____

_____

_____

_____

_____

_____

_____

OUT
TO
PASTURE

Chapter 10

The Pen

Pastured poultry require shelter, predator protection and efficient feeding and watering systems. In addition, the fresher the pasture, the greater the quantity the birds will consume and healthier they will be. Just pasture is not enough. It must be fresh.

Birds are extremely hard on pasture, for a couple of reasons. One is that they physically scratch the crowns of the plants, which is good but only for a short time. The second is that their manure is extremely high in available, soluble nitrogen. This "hot" manure, as it is called, can easily cause nitrate toxicity in the soil or water. An excess can quickly deplete humus in the soil just like chemical fertilizers. Capitalizing on this valuable resource, however, is an important goal. It should be assimilated into growing plants as quickly and efficiently as possible.

The classic picture of pastured poultry is the proverbial clay chicken yard. Traditional American homesteads had them. The chickens scratched out all the grass and their manure buildup turned the soil into brick. Wet weather produced a mudhole and dry weather produced a dustbowl.

One solution, of course, is to simply let the chickens free range around the homestead. This is fine for a dozen chickens, as long as the farmer doesn't mind chickens roosting on the back porch or in the equipment shed or eating vegetables out of the garden. But for commercial production this is not a solution.

High density, short duration grazing is the answer with poultry, just like it is with livestock. But chickens do not herd easily and they require shelter, protection and feed in addition to the forage. They are not herbivores.

Typically, poultry producers range birds on a large pasture for a relatively long period of time until the pasture is denuded of vegetative cover. Moving the birds is a major problem, not to mention negotiating equipment through a flock of hundreds to move feeders, shelters and waterers without panicking the birds or crushing some of them.

All these problems came together for us with a single solution, which we call the portable pen. This is a floorless cage, 10 feet x 12 feet x 2 feet high and light enough to be moved by hand with the aid of a little dolly. By breaking up the flock of birds into many small units, absolute control over pasture, feed and water can be maintained. The small groups mimic natural flock numbers, reducing stress and increasing efficiency. Each pen holds 75-100 birds.

Made of pressure treated softwood lumber to resist rotting, these pens weigh only about 150 pounds and can be moved with the dolly, which serves as a portable axle. Moving each pen takes roughly 20 or 30 seconds, and the birds walk on the ground as the trailing edge of the pen pushes them forward. Of course, they are eager to get to fresh pasture and therefore tend to stay well ahead of the trailing edge. Daily or twice daily moves ensure fresh pasture, not only to stimulate eating, but also for hygiene. Parasites and disease are a major problem of poultry under commercial production, but by giving them a new spot to lounge on every day, pathogenic problems can be virtually eliminated.

A framework of wood, carefully braced for strength, offers a cheap yet strong structure. The long pieces are pressure treated 1 inch x 6 inch boards ripped in half. These boards are ripped in thirds for all the braces. A 2 inch by 4 inch board, ripped in half, serves as the bottom

Photo 10-1. *The pen, 10 ft. x 12 ft. x 2 ft. high, made of pressure treated softwood, aluminum roofing and poultry netting.*

board on the front and rear of the pen, as well as the brace that holds the waterer. These pieces take quite a bit of abuse and need to be strong.

One end is entirely enclosed with aluminum roofing screwed into the light framework. We use screws throughout because of the flexing that occurs when we pick up and move the pen. Nails soon begin working loose. Using cordless drills and self-drilling, fine-threaded drywall screws allows us to build pens in a jiffy. The other end is one inch poultry netting on the vertical wall. Bigger netting allows predators to enter. Smaller netting, like hardware cloth, blocks out bugs like grasshoppers and crickets and is also more expensive.

The top of the non-enclosed end is made of two doors, one of netting and one of aluminum

Photo 10-2. *Note the turnbuckle and crosswire along the ground to keep the middle from sagging and the sides from bowing out.*

roofing.   The material is fastened to a light framework (heavier for the aluminum) that just lies on top of the pen framework. No hinges or fastenings are necessary, and in fact should not be used so that the doors can be completely removed from the pen to facilitate catching the birds on processing day. In windy areas like Nebraska and Kansas, fastenings may be required.

The totally enclosed end we run facing west, the normal direction of wind and weather. The east end is open to the morning sun. The netting door faces south to emit as much sunshine as possible. The opaque aluminum door covers the feed trough. The waterer, a commercially produced hanging type, is suspended from the top brace near the pen's middle, but under the netting quadrant door so that if it rains, the rain water will go into the waterer trough.

A 5 gallon white plastic bucket sits atop the pen on the eastern end, on the door and pen bracing, as a reservoir to gravity flow into the waterer. (See Photo 10-3.) The waterer and bucket ride with the pen when it is being moved. The trough feeder must be removed for moving. We use trough feeders because the long rectangle provides more linear feeding space as a ratio of feed volume than do round feeders. Bridging (feed not falling down) is also a major problem in round feeders.

In hot climates, it may be necessary to hinge the vertical wall roofing on the west end for increased ventilation. A Florida friend actually made the west end a peaked roof with netting in the triangular end to increase ventilation.

Another friend built a pen of aluminum conduit to lighten it, but it was so light that predators could come in underneath and a stiff

Photo 10-3. *Water reservoir and fittings to gravity feed hanging waterers.*

breeze could blow it over. He ended up having to stake it down every time he moved it. Another friend had the same problem (too light) after building one out of PVC pipe. The pen should be heavy enough to mash down grass.

Tarpaulins for the opaque shelter, rather than aluminum, can be used to heat the interior of the pen. But if the sun is up and the temperature is more than 55° F., the birds much prefer the cooler, reflective properties of the aluminum.

If the pen is built too heavy, it becomes cumbersome. Moving the pens with a tractor is unacceptable because it frightens the chickens. Movement speed is also much less flexible, resulting in injured birds. Hand moving is the simplest, quickest and most efficient way to go as far as I can tell.

A floored pen which the birds go into at night and free range out from during the day has many disadvantages, especially with Cornish Cross broilers, which are extremely lethargic. They simply will not walk enough to find and eat fresh pasture and bugs. They are bred to eat, drink, lie down and gain weight. The portable pens afford the birds fresh daily pasture without their really having to walk for it.

Although the birds do walk one pen length (12 feet), the rest of the day they are within normal breed movement parameters for water and feed. And their floor is forage. The portable house/free range model also requires much more building material. It is more disruptive of the birds' routine because they are penned up, especially until you get there in the morning to let them out. The bulk of the birds' eating occurs in the early morning, so it is crucial that that be the time they receive fresh pasture.

Predation during the day is a real problem. Raccoons, chicken hawks, foxes and other predators will attack in broad daylight. By keeping these chickens within the safety of the pen 24 hours a day, they receive the protection benefit along with the amenities of fresh pasture and lounge area. There is no building to clean out, no litter to haul. All the droppings go immediately on growing forage, which can utilize it right away to capitalize on the nitrogen injection.

The downfall of many alternative poultry operations has been with a chicken yard. In an effort to provide "range" to the birds, producers have not differentiated between stale range and fresh range. This is a critical factor because it spells the difference between success and failure. Many people view the pen as an unnecessary, labor-intensive component of this model, and balk at implementing it.

But you simply cannot get the level of health and forage utilization without the pen. If a stationary house with rotated pastures extending out like spokes is used, the areas closest to the house will invariably be overgrazed and turn to dirt while the far reaches are undergrazed. Pathogens build up around the house in the traffic areas and there is simply no way for nature to cleanse that area.

To my knowledge, nature sanitizes itself in only two ways: sunshine/rest, and decomposition. These limited alternatives have great import to livestock producers.

In the sunshine/rest mode, the sun, which is the great sterilizer, gently bathes the areas it touches with radiant heat that dries moisture. Ultraviolet radiation kills pathogenic organisms. The rest period is necessary to let this slow ray gun do its work. Most parasites, for example, cannot live more than two weeks without a host.

Most parasites, too, require moisture to go through their life cycle. Wet grass blades contain many more pathogens than dry ones. Realize that many of these critters are microscopic, and moisture provides the only way for them to paddle around from place to place. Deprive them of that, and they become fairly immobile - and localized.

The rest period is necessary to deprive these pathogens and parasites of a host. The rest period should include some dry time to help destroy the organisms, as well as be long enough to deprive parasites of a host.

Clearly this natural sanitation principle discredits any sustained-stay model that includes range.

But now let's move to the second natural sanitizer: decomposition. This not only occurs in the field, but is extremely important in housing models. This gets into the deep litter concept.

It is worth being reminded that as soon as we overload the soil or the house floor with too much nitrogen, we create harmful nitric acids, we destroy beneficial organisms, we lose nitrogen through vaporization or leaching, and we have a cold, hostile medium. Remember that the carbon/nitrogen ratio should be around 30:1, and that an ammonia smell indicates you need more carbon.

As limited as it is, confinement housing CAN be sanitary, but it requires a tremendous injection of management, energy and materials to keep it functioning within sanitation parameters. Our only confinement house is the brooder house, but that is only on the front end where not too much manure is generated, and not much costly housing is required. The big injections come as the birds get big, and then I am confident that the pen offers the most efficient alternative. Certainly a yard, even if it is rotated, does not do the job.

It is better to totally confine the animals in a well-lighted (many skylights), well-ventilated (leave the eaves open) facility than to confine them in a barn with continual access to an adjacent "exercise yard." Poultry, for example, that are classified "free range" on a yard that has constant poultry pressure, is hygienically unacceptable. The birds would be better off completely confined in a facility that meets the parameters of this discussion. In fact, some of the largest organic poultry operations in the country have succumbed to disease and gone belly up for exactly this reason. They had range, but not enough rest.

We all know the historical record of human plagues when hygiene wasn't followed and unsanitary conditions prevailed. Animals fall under the same rules. Disease outbreaks are not necessary. Any true husbandman can and will institute rearing models that maximize hygiene in his animals.

I cannot overemphasize the value of the pen to the smooth running of the pastured poultry model. It answers the need for small flock grouping, sanitary conditions, fresh forage, proper manure management and predator control. It is also far cheaper and simpler to maintain than a stationary house of equal square footage.

Certainly we would not want to disparage any attempt to properly pasture poultry, but for us the small portable pen seems to have addressed all the critical elements associated with commercial pastured Cornish Cross broiler production.

Chapter 11

Moving the Chicks
Out to Pasture

Sometime between two and four weeks, the baby chicks are ready to move out onto pasture. In extremely hot weather, they can be moved out earlier. In fact, we have moved them out at seven days, but I wouldn't advise that. It's too soon. The birds are simply too small to negotiate the grass. The earliest I would advise putting them out would be twelve days.

On the other hand, every week past four that the birds are kept in the brooder will cost you 10 percent in carcass weight at the end. Remember that the reason factory raised birds receive hormones and antibiotics is to compensate for the unnatural conditions under which they are being raised. Their air is full of fecal dust, their litter is filthy, they do not receive fresh air and sunshine, and they are not receiving green material or exercise. The brooder house period closely resembles conventional production and therefore must be cut as short as possible. If we raise birds without the artificials, hormones and antibiotics, we must necessarily expect tremendous disease problems as well as dramatic performance loss when they remain too long in the brooder house.

The only way these birds can perform well without negative inputs is to get them out on the positive pasture.

To minimize stress, move the birds early in the day. This has both the benefit of moving them when it is cool and of allowing them to have at least half a day in new surroundings before nightfall. It is hard for us to appreciate just how stressful this move is on the birds. It is far more stressful than a similar move when the birds are six or seven weeks old. Not only are the birds being moved to a new area, they are going from inside a house to outside in the pasture. It really does shake up their world. For that reason, we want it to go as smoothly and comfortably as possible.

Photo 11-1. *Rachel loads three-week old chicks into the pasture pens.*

Obviously you want to be aware of the weather report. You don't want to move the chicks out on a rainy day, or a cold, blustery day. If the weather is going to be nice for a couple of days, we try to get the birds out early. Or maybe we will move out half of them to try to minimize the risk. Moving out half at least gets some out and certainly takes pressure of those who remain. You can push it, or wait until the final hour, depending on what the weather is doing.

If a couple of good days are expected, I encourage you to go ahead and get the birds out. Within just a couple of days they acclimate well and take right off in the pasture. If the weather is good, the move seems to just stimulate them to grow faster. Once they are acclimated, even if it's just for a couple of days, they can take some rough weather. The critical thing is not to throw them into rough weather in the first 24 hours.

When you go in to catch the chicks, remove as many obstacles as possible. Take out the waterers, feeders, and get the lights and hover out of the way. Otherwise you'll be tripping and falling, or breaking things, and that gets both you and the birds all worked up. We use a couple of plywood pieces about 4 feet long and 2 feet high as partitions to crowd the birds into a corner. Don't try to get all the birds at once. Just ease some into a corner, pen them in with your handheld partitions, and then fill a crate.

Even though you can put more than 50 in a crate, especially if they are only 14 days old, do not. Give them plenty of room. If you put in 75 or 100, they will pile up and smother. It doesn't matter if it's cool or warm; when they get excited they will pile up. It's especially critical not to overload the crates when it's warm.

Be patient catching the chicks. Take two or three bites if necessary. You don't want to force more than 100 or so into a corner or they will pile up and suffocate. I can't stress enough the propensity of these little birds to pile up and suffocate. As they pile up in the corner where you are catching them, keep raking the top ones down to expose the lower ones to some air. They really do panic, and you must be aggressive about giving them air.

Do not be abusive when picking up these chicks, but you need not treat them too gingerly, either. I catch them by one leg, putting seven or eight in one hand and a couple in the other hand so I pick them up ten at a time. That makes counting easy. I slip the legs between my fingers to hold more. Be sure to pick the birds up randomly so you do not end up with a pen of cockerels and a pen of pullets.

After loading the crates, you will put them in a trailer or pickup to go to the field. The birds will spread out in the crate and calm down. Many of them will even go to sleep, sticking their heads clear through the slats of the crates. Be sure not to crush heads when you are loading crates. It happens real easy; I know.

Drive carefully out to the field so the birds are not shaking around. The more you shake them, the more they will pile up in fright. Just take your time; it really is worth it.

Have the pen situated where it is going to be before you put in the chicks. You want to give them at least 24 hours before they go through the trauma of their first pen move. Take the chicks out of the crates as fast as possible so you can shut the lid and leave them alone. If the move goes smoothly, within 20 minutes the birds will be up and exploring their new surroundings. Make sure they can reach the water and feed before leaving.

That first evening, make sure they are sleeping under roof. Check to make sure none has found a hole and gotten out. If all is well, enjoy your night's rest.

Chapter 12

Pasture Logistics

The pens should be arranged in a wing formation, with the lead pen on the downhill side so that in case of surface water runoff, the water running through one pen will not enter the adjacent pen. Another reason for the staggered arrangement is to enable the trailing pen to run immediately adjacent to the preceding pen's paddock. If the pens run side by side, alleyways for access must be left. This takes up space, and leaves a streaked appearance in the field when the grass greens up behind the pens. It just isn't as pretty or efficient.

We take water out in large trailer-mounted water tanks with large valves so that we can gravity fill 5 gallon buckets to carry water to the pens. These buckets may be any color. The pen buckets must be white. Black ones absorb solar energy and warm the water, reducing consumption. Yellow and blue buckets attract insects which die, fall to the bottom and then clog the small hose, couplings and water assembly. Without 5 gallon buckets our operation would fall apart. The world revolves around 5 gallon buckets.

Photo 12-1. *Underground storage tanks and old axles can be salvaged for water buggies. Rachel, our daughter, fills water buckets.*

Photo 12-2. *Joel fills pen water buckets.*

Feed is stored in large used fuel tanks. Underground fuel tanks can be had for the asking. I cut them up with an acetylene torch so they are no more than about 4 feet high at the rim. If you are afraid of an explosion, fill the tanks with water to displace any fumes that may be present. Clean them out with a wire brush and let rain into them for a month if possible. Marlin Burkholder in Singers Glen, VA, picked up several poultry shipping boxes from a large poultry processor. When the outer wall gets damaged, the company throws them away. These light double-walled plastic boxes hold about a ton of feed, and he moves them around with forks on his front end loader. When they are empty, they are light enough to move around by hand. The feed containers must be portable to move up and down

Photo 12-3. *Rachel dips feed out of a tank.*

Photo 12-4. *Joel scoops feed into the feed troughs after moving all the pens.*

Photo 12-5. *When the feed tanks are empty, we flip them over and roll them up the field to the next strategic location.*

the field with the pens. A gallon of feed weighs about 7 pounds, so you can compute how much feed a tank will hold by knowing the volume.

A simple top made of roofing, on a wooden frame that fits within the tank's circular rim, makes a fine flat roof. Normally enough slant in the field can be obtained for it to shed water nicely. This roof is light enough to be shoved aside or even lifted down and set aside while you scoop out feed with a shovel and put it in 5 gallon buckets for transport to the pens. I usually stand on a 5 gallon bucket to get some height advantage over the feed tank. A large galvanized feed scoop works well to dip the feed out of the 5 gallon bucket into the trough feeders at the pens. (See Photos 12-3 and 12-4.)

The bulk feed truck from the feed mill drives right out through the field and augers feed into our tanks in just a few minutes. When the tanks are empty, I simply flip them on their round sides and roll them down the field to where they will be close to the pens (Photo 12-5). If the ground is too soggy for the truck to get in when we need feed, we can put the tanks on a hay wagon and fill them from a solid road or lane, and then pull them out to the field with the tractor.

The pens should be moved one after another without feed and water being added. The fresh feed and water distract the chickens from their pasture. Delaying the feed and water a few minutes encourages the birds to eat more forage. After moving all the pens, I go back and fill all the waterers. That way I can glance at them for any malfunctions during the return trip, when I fill all the feed troughs and replace them in the pens. Diane Kaufman in Wisconsin uses 6-inch PVC pipe, cuts out one-third in the top, and hangs that in the pen for feeders.

The dolly we use to move the pens was invented by my engineer brother, Art. It is essentially a portable axle designed to function also as a lever with the wheels being the fulcrum. The 8-inch lawnmower wheels are not as prone to getting stuck in holes as small 6-inch wheels.

Made of half-inch pipe, it is light but strong enough to do the job. It allows me to move each pen by myself in less than 30 seconds. About 4 feet square, it is wide enough to give good pen stability but narrow enough so that the axle, which rides only a couple of inches above the ground, will not bind up on grass clumps. A refinement I put on the second generation was a bent axle, so that the pen rides the same distance off the ground - about 2-3 inches - but the middle of the axle is about 6 inches high. This allows it to clear the grass better and aids in going over

Photo 12-6. *The moving dolly.*

83

rough ground.

With the handle held vertically, put the prongs under the bottom board of the pen. Grasping the pen handle (a wire with a garden hose on it to keep from cutting your hand), step into the dolly so you are straddling it. Body position is important to make the whole movement easy. Pushing back and down on the dolly handle, and picking up on the pen handle, slip the dolly under the pen, and let the bottom board fit into the stop slot.

Go around to the other end of the pen and pull on that handle. The pen just rolls along. I usually ask one of the children to go out with me and help shoo the birds away from the trailing edge in the first week. It speeds up the process when the birds are tiny.

Photo 12-7. *Approach the pen with the dolly.*

Photo 12-8. *Slip the prongs under and grasp the pen handle.*

Photo 12-9. *Use the dolly as a lever to help lift the pen.*

Photo 12-10. *Ease the pen and dolly down. The pen is now ready to move.*

Photo 12-11. *Take the feeder out.*

86

Photo 12-12. *Go around to the other end of the pen and pull the handle. The chickens walk on the ground.*

Photo 12-13. *Daniel places plugs under the pen on extremely uneven ground. This keeps predators out and the birds in.*

87

Photo 12-14. *Chickens grazing after a move.*

Diane Kaufman uses a pair of roller skates to move her pens. She took off the shoe and fastened on a 2" x 4" board notched in the middle to keep the bottom board from slipping off. She slips one skate under one corner and the other under the other corner and claims it works great. She has 7 pens and moves them all by herself.

The daily maintenance per pen takes about 5 minutes. With feed and water tanks no more than a few yards from the pens, servicing is as easy as it is in a regular loose-housed chicken house. Moving the pens takes very little time. I can move and service 26 pens in 90 minutes: that's 2400 birds!

And the daily moves reduce feed consumption by 30 percent and produce a healthy, low-fat bird. The benefits far outweigh the labor involved in carrying feed and water and moving the pens.

With daily moves, and loaded with about 80 or 90 birds, each pen spot is putting down the equivalent of 300 pounds of nitrogen per acre. If we drop below 50 birds per pen, performance drops because the birds do too much running around. We are somewhat locked into the number per pen in order to maintain peak performance.

But 300 pounds of nitrogen per acre is a healthy dose in anybody's book. In fact, many agronomists would say it is too much. But that is why we want the birds to be on a fast-growing, vegetative forage. We want the forage to be able to assimilate and metabolize this shot of nitrogen. In a continuously grazed pasture, the forage becomes weak and can't respond when given a big meal. If you are working hard, are in good physical condition and burning off many calories, a big meal feels good. But if all you've done is laze around the house (like at Christmas time) and shuffle some papers, a big dinner does not sit well.

Grass that is in good condition, that has been kept at what we call energy equilibrium through intensive controlled grazing (short duration, high density) can utilize far more nutrients than weakened grass. This whole model involves many parts of a puzzle for a maximum payback. The more pieces you throw away as unimportant, the poorer the payback.

Healthy, vegetative grass can metabolize twice the nitrogen that continuously grazed, weakened grass can. Agronomists who say 300 pounds is toxic are not using properly managed grass in their study. After the initial flush of growth we come in behind the pens with the cattle and graze off the grass when it is only about 6 or 8 inches. I call this candy bar green grass. We run our portable electric fence at right angles to it so that the cows in any paddock have some of the candy bar grass and some regular forage. If all they get is candy bar grass, their manure is too washy.

One thing is for sure: cattle like that candy bar green grass. When cows on other farms are going through fescue toxicity problems, ours eat this fescue like candy. There is clearly no repugnance to the chicken manure. The cows take the grass right to the ground. As usual, I move the cows in 24 hours to another paddock, and let the just-grazed grass regrow. If soil moisture is good, this grass will often regrow nearly an inch per day. But this grazing pressure stimulates the grass to grow more than it would if we waited until it grew to 12 inches or more before grazing. If we left the cows on longer so that they kept nipping off the new shoots, the grass would be stunted and could not express, through plant volume, the benefits of the chicken manure application. By shutting off the expression through lack of a rest period, then, we would lose many of the nutrients through vaporization or leaching.

The idea is to keep the grass in its "blaze of growth" cycle that Andre Voisin explains in his classic, *Grass Productivity*, to fully utilize the nitrogen the chickens have put on the soil. It typically takes three grazings, or three shearings, to return the grass to normal green color. The intensive controlled grazing allows us to get the full benefit of the nitrogen, and also to fully harvest it rather than letting it leach or vaporize away.

How we place the pens and what we do, both in front of and behind the pens, play a crucial role in healthy birds, a healthy environment and a healthy payback.

Chapter 13

What Kind of Pasture?

Pasture consumption varies according to freshness, type and length of the forage. Certainly any forage is better than none, but ideally it should be the most palatable and the most resilient that can be offered.

Length is crucial for a number of reasons. Chickens like the forage to be somewhere between one and three inches. I've seen birds eat a 6 inch blade of grass, but that certainly is the exception, not the norm. They prefer those tiny, tender shoots about 2 inches long, both for palatability and harvesting ease.

This height can be obtained by mowing ahead of the moving pens or by grazing. Mowing leaves a stubble that seems to hurt the chickens' feet. They much prefer the sward condition after it has just been grazed by livestock. This grazing can immediately precede the chickens. No rest period is required because the residual stand contains plenty of forage to feed chickens.

The shorter the grass, the easier it is to move the pens. The trailing edge runs only about 2 inches above the soil level. If it ran higher, the chickens, especially when they are small,

Photo 13-1. *Cattle prepare the pasture for the chickens by shortening the grass and uncovering bugs.*

Photo 13-2. *Pasture before the chickens go on.*

Photo 13-3. *Pasture after the birds have been there for one day.*

Photo 13-4. *Pens move across the field like a flock of migratory geese.*

would slip out underneath it. When the sod is clumpy and/or tall, the bottom board snags and binds. The chickens move much better, too, when the grass is short. If it is tall and thick, they tend to get stuck against the pen's trailing edge as the grass bends over and traps them in a narrowing wedge.

The more difficult it is for the birds to run ahead onto the new pasture, the slower the pen must be moved to not run over them.

The hanging waterer, too, may malfunction if it touches a thick clump of grass. Some of the new models have a special independent suspension that allows them to touch the ground and still work. But many, especially older models, do not shut off if the base touches the ground. The float valve stays open if the waterer touches the ground. Water constantly running is as bad as water that won't run. If the valve won't shut off, the bucket reservoir soon runs dry and the birds are out of water until the bucket is refilled.

Tall grass, too, tends to reduce air flow around the pen, causing it to heat up inside on a warm day. It also makes large gaps between the pen's bottom board and the ground. This not only may allow the birds to escape, but also invites predator problems. These gaps, caused by soil depressions, can be easily spotted and plugged when the sward is short; they are hard to see when the grass is long.

Many predators tend to shy away from short grass. A great horned owl will pluck a weasel or mink right up off the ground at night. As a result, these predators like to walk along the edges of woods or through tall grass, which contains the greatest number of field mice and voles anyway. Mice and voles are generally the predators' preference.

Pasture length, finally, has a direct bearing on bird cleanliness. When the sward is short, the birds' excrement goes right onto the soil surface, where the birds scratch it in and the moisture is absorbed into the ground. Tall grass, however, bends over and forms a barrier between the chickens and the soil. Furthermore, this vegetative mat is slick, high in moisture, and similar to slick magazine paper in absorbency compared to Bounty paper towels.

In just an hour or so, the slick mat of grass becomes soiled and the birds slip and slide around on this until they are moved. For hygiene it is imperative to keep the sod short.

The pasture species should be as diverse as possible. Just like livestock, chickens will vary their diet according to season and availability. The birds love seeds. Ragweed, plantain spires and other weeds that produce abundant seeds are a favorite of the birds. Legumes are their next favorite, and then come the grasses. I hesitate to rank the species in order of preference because someone may then try to provide only the most desirable thing. Actually, in their first few minutes of grazing, some birds eat fescue and others eat seeds and others eat clover leaves, just as people would pick over a salad bar that would contain "favorites" and "I need to eat this because it's good for me" items.

I do not encourage producers to run the chickens on monocultures or annuals, simply because nature never feeds animals that way. Animals constantly forage for a diversity. To take the abuse that the chickens mete out, a perennial sod is best. A planted annual crop simply has too much soil between plants to offer good loafing conditions when the sward is only one to three inches high. A perennial polyculture offers diversity and vegetative density.

I would not suggest that chickens never be offered alfalfa pasture or cereal grains. But monocultures never provide as much cover or volume at the proper sward height. The perennial polyculture offers the best mix of all the right things for maximum pasture performance.

Every geographic region has its native forage species. I have not found any forages that the chickens dislike. Whether it is fescue or lovegrass, the height and density seem far more important. In areas where grass grows sparsely, it may be necessary to move the pen more frequently to ensure that the birds get enough to eat. The trade-off is, of course, that in such arid areas the grass is richer in vitamins and minerals so that not as much needs to be ingested to get the same benefit. Further, such areas would not have the wetness that plagues more temperate areas.

Remember that in cool conditions, bare soil tends to be colder than a dense sward, and in hot conditions a dense sward is cooler than a sparse sward with quite a bit of exposed soil between the forage tillers. This is another reason not to cultivate and plant annuals or monocultures for the chickens.

The point of all this is that forage species are just not that important. The critical factor is that it be fresh, short and preferably composed of many different species so that the birds have a great variety.

PROCESSING

Chapter 14

On-Farm Slaughter:
The Advantages

Pastured poultry is valueless until it is
prepared for cooking. That means the birds need
to lose their heads, feet, feathers and entrails.
Processing is what scares off many would-be
poultry graziers. Poultry processing differs
markedly from beef, swine, sheep and goat
processing.

Poultry processing in America is concentrated in the hands of a few large vertically
integrated poultry companies. These companies
own their own breeder flocks, feed mills,
hatcheries and processing facilities. Not many
years ago, it was common for each community to
have a small poultry processing facility similar
to the many neighborhood livestock slaughtering
facilities still in existence. But with the
integrators these facilities have largely gone by
the wayside.

To my knowledge, there is only one in
Virginia. States do vary significantly on what
is allowable and available. For example, some
allow state inspected facilities to process and
sell those birds anywhere within that state.
Still others have several small federally in-

spected facilities. Prices vary from 60 cents a bird to $1.50 or more. In any case, in no state are these facilities as accessible or numerous as the local livestock slaughtering facilities.

Most producers must drive 20 miles or more to get to one and then pay an average of $1.00 to $1.25 per bird for the processing. Transportation exacts a heavy toll on these birds due to stress. Remember that their metabolism is extremely high. Being deprived of feed and water for a couple of hours is equivalent to a beef going a day without nutriment. The animal tightens its muscles and begins shutting down vital functions to conserve body moisture and stay alive.

Animal rights activists, for all their misdirection, are right on target when pushing for animal slaughter as close to the point of production as possible. Not only does it relieve stress, a direct cause of tough meat, but it is far more environmentally sensible. The cost of trucking the live birds, then processing the offal, then transporting the carcasses across the country is much higher than a more local or regional slaughtering approach. Processing on the farm where the birds are produced is best.

But the reason for on-farm slaughtering goes far beyond both stress and logistics. The way large scale processors slaughter birds can render the best pastured poultry unfit for human consumption. The media are being flooded with reports regarding *Salmonella* contamination and unsanitary practices in the poultry processing industry. Every time one of these hits the press, we receive calls from people wanting to bail out of the conventional poultry market.

Just for examples, let me describe a couple of procedures that are crucial. The first is

execution.   Conventional processing electrocutes the birds, then slits the throat. This
procedure reduces muscle contractions and reduces the ability of the carcass to bleed. The
Biblical Mosaic law of slitting the throat is by
far and away the best way to kill animals, not
electrocution. We put the birds in a cone and slit
the throat so that the autonomic nervous system
keeps functioning and the heart actually pumps
the blood right out of the body.

Evisceration is done, conventionally, with
a stainless steel loop that exerts pressure as it
scoops out the entrails, or by vacuum. Often the
intestines tear apart and the feces splash all
over the inside and outside of the bird,
contaminating the meat. Some processors put the
birds through as many as 40 separate chlorine
water baths to kill the germs caused by all this
fecal contamination.

The large refrigerated cooling vats in which
the birds are placed after processing accumulate
as much as a foot of fecal sludge in the bottom
after a day of processing. In fact, some reports
say that 10 percent of broiler weight in the meat
counter is composed of this fecal soup. The
muscle tissue, lacking tone due to production
methods, is highly absorptive and soaks up this
fecal soup.

The point of all this is that the production
model merits nothing unless it is consummated
with a complementary processing model. A very
small processing facility can stay much cleaner
than big ones. For example, we only process about
4 days every 3 weeks. In between times the sun
sterilizes our shed, rains wash away bits of blood
and feathers, and anything else gets eaten or
decomposed by microscopic critters. We eviscerate by hand so we don't tear the intestines and

rarely spill feces anywhere on the bird.

Checking over the birds carefully to make sure they are squeaky clean keeps the water in the cooling vats extremely clear and clean. The fact that our customers come right into the processing shed and see their chickens in the cooling vats requires that we run a clean ship. We have customers who work in the processing plants but purchase their chicken from us, vowing that they won't eat one out of the store.

We have customers who have allergic reactions to chemical residues and chlorine contamination on normal birds, and yet have no problem eating and handling ours. The fact is that it is much easier for a person to keep his home kitchen clean when it is used to prepare 10 meals per day than if it were used for 1000 meals. The magnitude of the processing facility inherently changes cleanliness and control.

By slaughtering on the farm, the chickens do not undergo transport stress, and we need not haul them away. We get to stay here on the farm, which is where we like to be. We process in the morning and customers begin arriving at 1 p.m., bringing their ice and coolers with them. We need not refrigerate, bag or deliver. We do what we do best - produce the world's best, cleanest chicken - and our customers meet us halfway. We can't bag, deliver or freeze any better than conventional services. By focusing on the niche in which we excel, we capitalize on our strengths and do not get bogged down in ancillary services. We have no desire to feed the world or build an empire. The small, efficient on-farm production and processing encourages regional food self-sufficiency and sensible marketing.

For people who have dressed poultry by hand,

the idea of a husband and wife processing 50 birds an hour or better is practically inconceivable. But it's like any acquired skill. After you do it enough, you get good at it. The automation then augments your own skills and allows you to go even faster. Much is technique, much is organization, and much is just physical strength and endurance.

From an income point of view, though, let me illustrate why this do-it-yourself processing is important for the success of your venture. The average cost of processing a chicken is $1.25 per bird. If we add the transportation time and expense, assuming an average of 50 miles round trip, the cost goes on up to $1.50 per bird. Teresa and I can do 50 an hour, which is a saved value of $75. Divided in half, this would figure out to be $37.50 per hour for each of us. Taking off a couple of dollars for depreciation on equipment and energy costs, we are left with about $35 per hour for our labor.

I don't know about you, but I have no problem working for that price. I can do just about anything for a few hours a month to earn that kind of money. I guarantee you no contract poultry grower working for one of the large integrators is making that kind of money per hour. Most of them are happy to make enough to cover the mortgage on the confinement facility, the fuel bill and maintenance costs.

All we have dealt with here is the economics of the decision to process on the farm. We haven't addressed the quality of life issue, of being able to stay on the farm with your work instead of going elsewhere, and fighting traffic. When you take the whole picture into consideration, the on-farm processing is the alternative that yields the greatest return in both quantifiable and nonquantifiable measurements.

Of course, if anyone comes by to help, the processing goes faster. If Mom helps (and she usually does), we can speed on up to 60 an hour and if Daniel, our 11-year-old, helps, we go on up to 70 or better. Many things can by done by children just as efficiently as by an adult.

We have customers who occasionally like to come out and help just for the fun of getting "connected" to their food. Wouldn't it be wonderful if all the energy uselessly expended in walkathons and similar fund-raising activities, could be channeled into farm work? But when we transport the processing off the farm, we can't utilize that energy. Furthermore, folks who want to learn or teach their children where their food comes from miss the opportunity. If one of our ultimate goals is to reconnect the urban and rural sectors of our culture, on-farm processing affords us a technique to accomplish that goal.

About half of our processing days now we have some sort of help, whether it is a customer or someone coming to learn the skill and then begin raising chickens themselves. If we love people, and make room for them, we can give them the pleasure of being helpful in many ways. If someone does not want to get involved in the processing, we can use them carrying feed or water to the pens, which expedites chore time and allows us to get back to the processing faster.

The point of this whole discussion is that when you are at home you can teach, you can utilize free labor from those who yearn for a food connection in their lives, and you can save a pile of money in the process.

Invariably, someone will look at our labor return and surmise that hiring the work done would be ideal. Why is it that the goal of most

Americans is to never get their hands dirty again? Our culture has for the past generation succeeded in reducing honorable work to something akin to criminal activity.

Hard work is wonderful. It allows you to enjoy a big meal without getting fat. If we would spend all the energy we expend on exercise equipment and playing at the local fitness center on meaningful, calorie-intensive work, our economy would be the envy of the world. We would be healthier, happier and richer. But no. Physical exertion is supposed to be shunned like a plague, and the ultimate "job" is the one that offers the most vacation, the most relaxation and the least exertion. One of the most frustrating experiences I have is when college students come, especially those taking agricultural economics courses; within five minutes they are telling me that all I need to do is hire all this work done so I can go in and sit at the desk all day. I don't want to sit at the desk all day.

Hiring people exacts a toll on your quality of life and quality of your product. An employee does not have the same stake in the success of the business as you do. He is not as committed to each detail being done properly. You simply cannot duplicate yourself. Sure, we hire people occasionally to do some things, but it is extremely limited and the job must be closely supervised. Hiring the processing done with part time committed labor can be done, but I think that needs to come later as the business grows. Initially I think you need to do it yourself.

An employee will not turn out the quality at the same speed that you can. Besides, as soon as you have an employee, then you have a host of paperwork to fill out. You must pay workmen's compensation, you have withholding and other things to deal with. It is not a free lunch, for sure.

No, I still say the best way is to use family; get your own hands dirty so you can see firsthand what those chickens look like. There is no substitute for keeping your hands right on the pulse of the product. The benefits and personal satisfaction of processing the birds yourself, ensuring the quality right through to the customers, being personally responsible and gratified, makes on-farm processing the way to go.

One advantage worth noting is in the disposal of processing water and slaughter wastes. That is a major cost for the commercial poultry industry. A plant that processes 100,000 birds per day literally has tractor trailer loads of offal, feathers, blood and millions of gallons of processing water to treat.

We use about half the water per bird required in the commercial processing facilities. They use about 5 gallons per bird and we use only about 2.5. That means right off the bat we have far less usage and far less effluent.

Because we only process a few days per month, and use only a few hundred gallons at a time, and our processing is in the summer when the soil is dry and plants are growing, we can use our water for irrigating the garden or run it out on the pasture, and we do not overload the environment's ability to break it down and handle it. As soon as you begin routinely dumping thousands of gallons of water on an area, and do it off season to boot, a whole new cost in treating the effluent becomes necessary.

That is one reason why I am not interested

in a central facility for processing the birds for several pastured poultry producers. I would much rather see a portable rig that could go from farm to farm and spread out the effluent and offal in such a way that the environment could handle it without expensive treatment facilities. It is also a reason to stay relatively small, rather than trying to build an empire.

By composting, the offal becomes a wonderful asset rather than a liability. And how much simpler to handle it through composting a few feet away from the facility than trucking, rendering, dehydrating, grinding and all of the things entailed in handling it by the commercial industry.

From every angle, then, on-farm small-scale processing offers the best alternative.

Chapter 15

Slaughter Mechanics

Twelve hours before slaughter I remove all feed to let the craw clean out and make processing easier and cleaner.

Using a plywood partition like a giant paddle, I push the chickens to the front of the pen. By removing both doors, I can catch the chickens easily and put them in crates for transport to the processing facility. I do not put them in a large container or they will pile up and suffocate. It is important to use crates.

On the following three pages, Photos 15-1 through 15-6 illustrate how we catch the chickens on processing day.

Photo 15-1. *Catching chickens on processing day. Remove the feeder and solid door and insert the partition.*

Photo 15-2. *Use the partition as a paddle to ease the birds around the front of the pen.*

Photo 15-3. *Finish the partition.*

Photo 15-4. *Set off the other door. The birds won't jump out.*

Photo 15-5. *Catch the birds and load the crates. We usually load 8 to a crate.*

Photo 15-6. *Loaded and ready to head to the house.*

To kill a chicken, we put it head first in killing cones. The head sticks out the bottom. The goal is to slit the main artery with a knife, but not cut the windpipe. This way the bird dies slowly and does not go into shock. The autonomic nervous system continues to function, allowing the heart to keep pumping and actually pump the blood out of the bird.

The cones can be mounted over a trough which ducts the blood into a bucket. The cones eliminate bruising when the muscular contractions occur, and allow control over the blood.

Photo 15-7. *The "disassembly" line and processing shed. In the rear, Joel loads birds in the killing cones. Then comes the scalder and picker. Daniel, our son, works on the eviscerating table and Mom, Joel's mother, checks things over on the "quality control" table. Initial soak is in the double sink beyond her elbow, with final vats beyond on the floor. A concrete slab was poured around the locust poles to build the facility.*

Photo 15-8. *Joel loads the birds in the killing cones.*

Photo 15-9. *Slit the throat. Blood drains into bucket at end of trough.*

Photo 15-10. *Hanging the birds on our homemade shackles in the scalder/dunker.*

Photo 15-11. *Up and down motion is essential for a good scald.*

114

After contractions have ended, the birds need to be scalded to loosen the feathers. The procedure we like best is about 145°F for about a minute. It is helpful to pull the birds clear out of the water several times to encourage feather flocculation, which helps displace oxygen at the feather follicle and aids water penetration. We add a couple tablespoons of Shaklee Basic H biodegradable soap to the water to break surface tension and encourage penetration. The soap also cleanses the birds so that they come out squeaky clean.

I made a set of shackles from boards and high tension band steel that encircles bundles of fence posts or packing crates. Some people use a basket. Anything is fine, as long as the birds move in and out of the water at a certain temperature for a certain period of time. Water access to every square inch of the bird is essential. The more uniform it is, the more uniform the scald.

Scalders and dunkers can be purchased commercially. We made our scalder by building a wooden box inside a box. Then we purchased a submersible thermostat and a couple of 4500 watt hot water heater elements and took the box and parts to a sheet metal company. The fabricator covered the exterior with a heavy gauge galvanized sheet metal and fit a tank inside, soldering the joints so it would hold water. He plumbed in the thermostat about halfway down and put in the elements through the wall about 8 inches up from the bottom of the tank. The stale air space between the two box frames serves as insulation. The dimensions of the water tank are 22 inches x 17 inches x 30 inches deep. It holds about 35 gallons of water and will keep water hot for roughly 700 pounds of birds per hour, which is equivalent to 125 chickens. Scalders are always rated in pounds per hour.

The dunker is a counterbalanced piece of pipe rigged to a geared down electric motor that gives us about 8 cycles per minute. Although this system may not be as snazzy as a factory-made outfit, it is far less expensive and quite acceptable for processing at our speed.

The idea of the scalder/dunker is to keep the birds moving in the water, preferably in and out, because this movement is essential to getting good water penetration. Commercial outfits use a shower gauntlet and high pressure water. If the birds just sit in the scald water, the difficult feathers simply will not loosen by the time the tender areas like the breast and the thin skin between the thigh and breast begins to burn. The action ensures that all parts of the bird get equal water access.

Our dunker automatically turns off and on. It has both a timer and a button, or switch, that the dunker arm depresses when the time runs out. In order to shut off, the assembly must be both out of time and depressing the button. That ensures that the chickens always stop outside the scalder. If we just had it on a timer, the birds would often stop in the scalder and I would have to flick the switch on again for a few seconds to raise them up so I could get them off the shackles.

I originally made this assembly so that Daniel, who was then only 5 years old, could dunk the four birds at a time. It worked so smoothly that immediately I asked an engineer to design a system that would be fully automatic. It helps to have smart friends. You can be sure that Bill Anderson gets all the free chickens he wants.

If you are buying new components for this assembly, I think an 8 cycle per minute motor would be fine, and you would want to rig it up like a steam engine. Come off with a cam and straight arm. By counterbalancing the dunker assembly, you could get by with only about 5 pounds of push or pull, allowing you to utilize a small motor.

Of course, for $4,000 you could buy a state-of-the-art automatic, stainless steel job. I would encourage you to stick with the Ford until you can afford the Cadillac. With our jury-rigged thing we can go 100 birds an hour. That should be enough for awhile.

From the scalder/dunker, the birds go into the automatic feather picker. All of this equipment is available commercially, and a set-up to process up to a couple of hundred birds per hour can be acquired for $10-12,000. The expensive components are scalders and pickers. Our picker will do 4 chickens in about 15 seconds. It is rated for 400 per hour.

The secret of a good pick is the quality of the scald. If the scald is too hot, the skin will tear. If it is too cold, the feathers won't loosen up. A cold water spray during the picking helps the birds come out cleaner.

After the picker comes evisceration. We pull off the heads. Cutting takes far more time and leaves little bone shards at the end of the neck. Pulling ensures neck severance at a vertebral joint and makes a clean job. then we cut off the legs.

We leave the necks on the birds. They add weight to the carcass, add material to the broth when the bird is cooked, and give us a good handle for picking up the birds and letting the water

Photo 15-12. *The automatic feather picker does 4 birds in less than half a minute. A cold dash of water keeps it clean and helps the feathers come out.*

Photo 15-13. *Pull the heads off. No bone shards that way.*

drain out. The broth from meat around bones is best, so whatever broth comes from the neck adds a high quality to the stock. If you take the necks off, I recommend using a heavy meat cleaver. We have only had one customer who wants her necks off, and we just put them in a separate bag and weigh them along with the birds just like we do the hearts and livers. All the necks can be cooked together with other bony parts to produce a wonderful broth. Our customers know that when they order a chicken, they get a whole chicken, including neck and tail.

Evisceration is like shearing a sheep. Experts make it look effortless, and it is fairly easy, but to the novice it can seem impossible. Technique is extremely important.

I start by cutting out the oil gland, on the top of the tail, by supporting the tail with the index finger of my left hand and using the knife with my right hand (I'm right-handed).

Flipping the bird on its back, with the neck facing my right hand, I pinch the skin with my left hand and with my right hand barely slit the skin between the neck and the breast, tearing open the skin and pulling the craw, esophagus and windpipe loose. If the bird has had nothing to eat for a few hours, the crop, or craw, will be empty and flat. That makes it a little harder to grasp and pull loose, but makes the whole evisceration process much cleaner since it can be pulled through the breast and out the back. If it is full, I cut it loose close to the breast, which allows digestive juices to come out. Birds that we dress before daylight do not get fed the prior evening. Birds that will be processed after daylight do not get fed that morning. In either case, time without food is essentially at night when the birds are inactive anyway.

Photo 15-14. *Cut off the legs.*

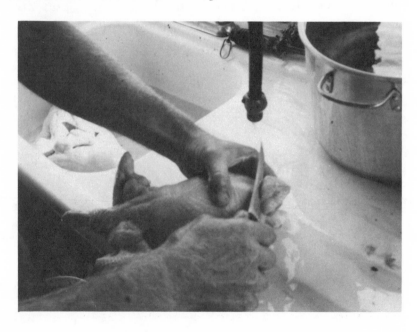

Photo 15-15. *Cut out the oil sack on top of the tail.*

Photo 15-16. *Slit the skin to dislodge the craw.*

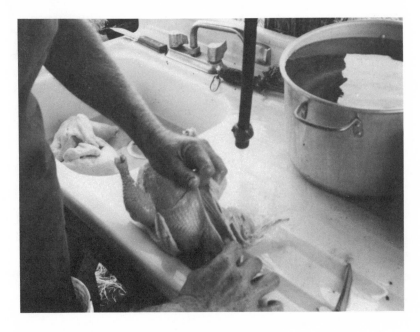

Photo 15-17. *Pull open the skin to expose the craw.*

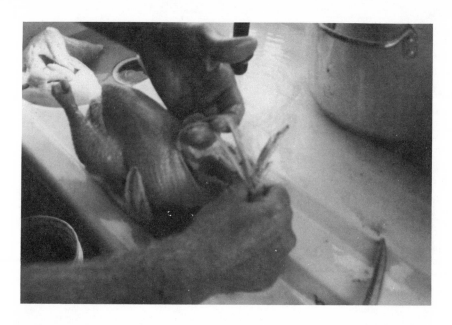

Photo 15-18. *Dislodge the craw and pull the esophagus and windpipe out of the neck.*

Photo 15-19. *Slit the skin just above the vent.*

122

Photo 15-20. *Pull open the rear enough to get your hand in.*

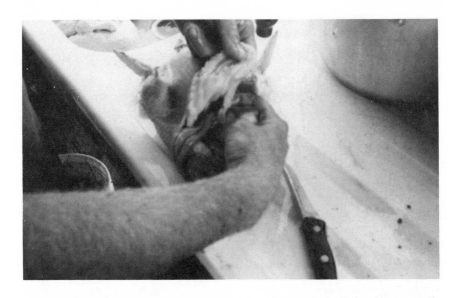

Photo 15-21. *Salvage all the fat you can and then slide your hand in, keeping your fingernails up against the smooth keel bone.*

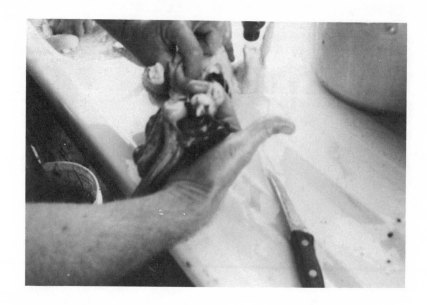

Photo 15-22. *Craw, windpipe and esophagus come out with all the offal.*

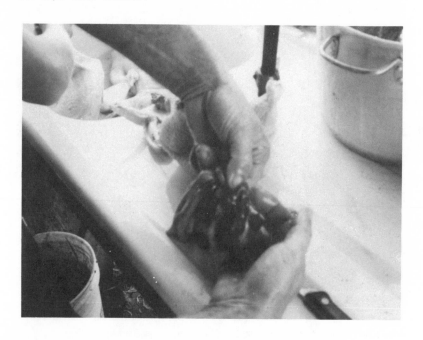

Photo 15-23. *Salvage the heart and liver.*

Photo 15-24. *Clear the offal until you're down to the single large intestine. Hold it to one side and cut down next to the vent, then repeat the procedure on the other side to cut the vent out.*

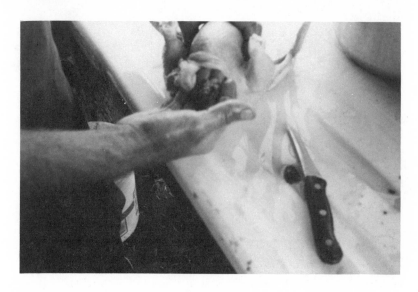

Photo 15-25. *Make sure the lungs are out. Then squirt water in and flush out the body cavity.*

Photo 15-26. *Slit the skin above the rear.*

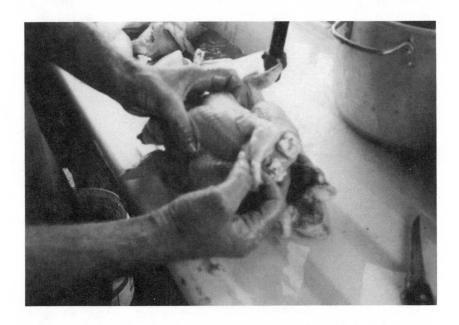

Photo 15-27. *Tuck in the legs for a compact, pretty carcass.*

After loosening the craw, windpipe and esophagus, I flip the bird around and with the left hand, pull up on the loose skin between the keel bone and the vent while cutting an incision just above the pubic bones, horizontally.

With both hands, I tear that rear opening bigger until I can get my right hand in. Pushing my hand in, I run my fingernails up high against the smooth keel bone. I run it way to the front, where I can loop my middle finger around the esophagus. Scooping down with my finger and pulling my hand back, I pull everything out but the lungs.

Then I take off the heart and liver. I pinch off the green gallbladder (attached to the liver) with the thumb and index finger and remove the gizzard if we want to keep it. It has two tubes coming out of it. One goes to the intestines and the other connects to the craw. The intestine tube can be jerked loose; the other one is cut with a knife.

The only thing attaching the offal to the chicken at this point should be a solitary large intestine, running straight to the vent. Holding slight tension on the offal, I take a knife and cut a U around the vent. Everything falls free. Then I reach back in and scoop deep into the rib cage to pull out the lungs, first one side and then the other, and check for the windpipe to make sure I got all of it. Often a piece will stick up to the breast, which is the roof of the body cavity when the bird is on its back.

I squirt water inside the body cavity to wash out any pieces and the job is done. For appearance, we like to cut a little slit in the rear, below the keel bone, and tuck the legs into that loose flap of skin. That tends to make the bird look more plump and compact and also helps it package nicer when the customer wants to put it in a bag.

The birds then go in a cold well water bath which takes away the balance of the body temperature and soaks out any stray bits of blood. After about 20 or 30 minutes, we shift the birds to a large cold water vat, where they stay for a couple of hours until customers come to pick them up.

The total time from execution to initial cold water soak is less than 10 minutes. A good eviscerator can gut nearly 100 birds per hour. It is a skill which improves with practice. With this setup and our procedures, a couple can process about 50 birds per hour.

Individual rubber garden hoses with plastic valves give each person his own water at the preferred volume. A 12-inch hose section after the valve keeps the water from spraying. Waste water from working tables and vats drains into a plastic pipe, which ducts it out to the garden or pasture. This nutrient-laden waste water irrigates growing plants. The volume and frequency of our waste water is easily assimilated by the plants.

Our processing routine goes like this. I get up at 4 a.m. and switch on the scalder. We get up a few minutes before 5 to start processing the 150 or so birds I caught the previous evening before dark and parked at the processing shed. We finish those around 7:30 and drain the scalder. After rinsing it out, we refill it and plug it back in. The cleaner the scalding water, the better the scald. Teresa slices open the hearts to push out clotted blood. The hearts and livers come out of the bird attached, and we leave them that way.

The hearts often contain a fair amount of clotted blood, which comes out easily when the heart is sliced in half and spread open. We leave the hearts and livers in one container.

We used to clean and sell the gizzards with the birds, too, but stopped that several years ago. We found that people didn't want the gizzards and even with an automatic gizzard peeler, there just wasn't enough money in the gizzards to justify the labor to clean them. Our time was much more profitably spent throwing away the gizzards and dressing a couple more chickens in the amount of time it would take to clean several hundred gizzards. Some people do like gizzards, though. If a customer really wants the gizzards, we save them but don't clean them. We give them away to folks who want to clean them themselves. Otherwise, we just throw them in with the offal.

I leave for the field and do chores, moving and servicing the chickens and catching the other 150 birds. Teresa gets breakfast for the children, organizes supper and cleans the hearts.

When I return from the field, the scald water is hot and we are ready to start the second go. I usually grab a couple of muffins or something and drink a glass of milk before we begin again between 9 and 9:30. We like to finish the second 150 birds between noon and 12:30, which gives us at least half an hour to clean up before customers begin arriving at 1:00. We clean down the blood trough under the cones, clean the scalder, shovel up the feathers and carry all the offal out to the compost pile.

We have chairs around so early arrivals can sit and wait for us until we are ready to serve customers at 1:00. We pick the chickens out of the vats randomly and try to drain out as much water as possible. Picking them up by the neck facilitates this. We have a set of platform scales calibrated to a big washtub so we can weigh about 25 birds at a time.

We put the number of hearts and livers in a bag to correspond with the number of chickens the person is getting. If they don't want the hearts and livers we leave them for someone who wants extra. We use a whiteboard to communicate that sort of thing to the customers.

If the weather is extremely hot, we freeze some buckets of ice in the freezer overnight and drop those big ice chunks in the vats to help keep the water cool. Since we move the birds so fast, it is not necessary to actually refrigerate them. Our well water comes out a little over 50°F and that is plenty cool to harden the fat. We want the birds cool enough to keep the fat hard.

By 5 p.m. all the customers have come and the birds are all gone. No bagging, no transportation, no cutting up, no freezing. It's all simple and low-key.

No-shows have not been a problem. Once in awhile someone will forget or call with car trouble, but our customers know that chronic problem-makers get excised from our customer list. They need to get with the program or they just don't have the privilege of getting our chickens. With an extenuating circumstance, of course, we are more than happy to work with a customer. But as a policy, we expect them to be here when they say they will be. If we have to freeze or bag birds due to negligence of the customer, we usually add on a hefty charge. If they have a good attitude, it doesn't bother them because they realize the hassle they've put us to.

If they are incensed about the charge, we may lose a customer. But it's far better to have a few appreciative customers than a lot of customers, half of whom are a pain.

We do not consider our afternoons with the customers to be real work. It is the most enjoyable time for us, and often folks just sit around for awhile and visit. It's payday for us, and there's something about seeing all those checks coming in that gives us a shot of adrenalin and helps us keep smiles on our faces even when we've been up since 5 butchering chickens. The continual stream of customer compliments is just icing on the cake.

Chapter 16

Composting Slaughter Wastes

We compost the offal with wood chips. A layer of chips, then a layer of offal, including feathers, makes a rough compost. We let the material heat. After it cools off, we use it for the compost carbon base a second time to fully utilize the chips and make a better compost. Several wire mesh gates hold the pile sides up vertically and aid in pile building. The heat keeps varmints away. The compost is a wonderful fertilizer for field or garden. Composting is superior to burial, incineration or rendering.

Composting can be done with any carbonaceous material, but it is helpful to have a friable type like leaves, sawdust or wood chips, as opposed to hay or straw. As you can imagine, the offal is heavy and sloppy. Hay and straw do not mix well with it and tend to form vent holes, which duct odors to the surface of the pile.

We use 3-foot high gates along the edges to keep the pile upright. Again, the sloppy offal and friable carbon do not stay up in a nice straight-sided pile by themselves. We keep a fringe of carbon about a foot thick around the edges as a buffer between the offal and the outside of the pile. Not only does this keep varmints from getting into it along the edges, but it also helps contain odors.

To start the pile, I put up the gates about 6 feet apart and put a gate across one end. Then I put down about a foot of chips: I don't want to lose the compost into the ground. Then I put on feathers about 3 inches thick. Obviously, there will be globs that are thicker and some areas thinner, but the general idea is to layer the feathers on about 3 inches thick. Then I cover them with an inch or two of chips before putting on the guts. I spread out the guts as evenly as possible, remembering to leave a foot of buffer all the way around the pile. The guts should be covered deeper with chips than the feathers. The rule of thumb is about an inch or two of guts and 4-5 inches of chips. The blood and cone-cleaning water goes into the pile, too. I spray down the top with water when I finish to help seal up vent holes. The next day's offal can go right on top until I get the pile up as high as I can conveniently toss the material.

At that point, I place two more gates 6 feet apart and butting up to the previous two, elongating the pile sides. Then I begin building between those gates. I always build the pile high and then elongate it in steps, rather than beginning long and building it a few inches higher at a time. It takes a critical mass of about 3 feet on all sides to generate enough heat to begin the composting process. A long low pile will not have enough height to get the compost started.

Overnight the pile should begin heating, and it generally heats for a couple of months. Three weeks later, when we butcher again, I usually pour a few buckets of processing water over the old pile to stimulate heat. The offal contains plenty of moisture to get the composting started, but unless the season is extremely rainy, it usually shuts down due to lack of moisture. A few inches of soil or foot of hay or straw on top as a cap will help preserve moisture and lengthen the heating period.

I am becoming increasingly pleased with the response from biodynamic preparations in the pile. Though not biodynamic practitioners, we have been toying with the preparations in compost for several years under the gentle persuasion and tutelage of Hugh Courtney. My observations suggest that the preps reduce odors and also prolong the active composting process by reducing moisture loss and heat excesses. Perhaps the day will come when I am a true-blue practitioner. I think biodynamics is in the refinement category, not the initial "necessity" category.

I reuse the compost to rebuild the pile with another injection of offal late in the season or the following year. Because of the offal qualities - smelly and soupy - we generally do not get a real good carbon breakdown during the first composting cycle. By sending it through another cycle, we get full carbon utilization and make a much better material. We do not turn the pile except in the recomposting stage. Rather than try to hurry the process, we let it age slowly and spread it the following year. Everything but the bone composts. Feathers do not compost readily, but given enough time they do fairly well. Generally, we recompost the pile with beef offal. We slaughter our beeves in October or November, which is timed perfectly to reuse the old chicken offal compost pile. Having gone through one cycle, the material fits in around the huge stomachs, the 100-pound paunches, of the beeves. Of course, we put in hooves and skulls as well as intestines and offal. By Spring, this combination of double-composted material, using two different animal materials, is truly a remarkable soil builder.

It takes about three or four cubic yards of carbon for every 1,000 birds, so we plan ahead and stockpile plenty of material. It is better to use a little too much than not enough.

Burying the offal entices varmints to come in and dig them up. Our experience is that the material does not decompose readily and stays gummy and slick for well over a year in the soil. It emits a horrible odor, too. Dogs and 'possums discover it and dig around, releasing offensive odors through the digging holes.

The compost does not smell offensive like you would think it might. We do all our composting within a few feet of the processing shed, which is where customers come to pick up their birds. Obviously if it smelled bad we would not put it in such proximity to the traffic area. Again, we use it as an educational tool. Often customers come over while I'm building the pile and learn about composting. The discussion inevitably turns to what they can do in their own backyard. That information transfer is just another one of those neat things that occurs when consumer and producer team up in a mutually beneficial relationship.

On a real hot sunny day flies can be a problem. We sprinkle some Basic H around and spray the floor and surrounding grass with the soapy water. That does a marvelous nontoxic job of keeping flies away.

Chapter 17

Inspection

We are inspected with far higher standards than government inspection: **CUSTOMER INSPEC-TION.** Our customers see the birds processed and there is nothing to hide and nowhere to hide it. They see everything. We have an open door policy, no trade secrets, no processing secrets.

Because of that, we must run a clean ship. Government inspection does not ensure food safety. Clothing a person in a white coat and giving him a title does not make him honest or fair, any more than putting a backwards collar on a man makes him righteous. Consumers must take responsibility for their own food safety. Delegating that responsibility unnecessarily is crazy.

It is amazing how many people who murmur about the corruption in government, how taxes are wasted, how fraudulent bureaucrats milk public programs and policies, somehow think that this same entity can suddenly do a commendable and laudable job as guardians of the nation' food. What schizophrenic reasoning! It is unreasonable to delegate food safety to government, in light of its reputation. Consumer scrutiny, individually or shared, is most efficacious for ensuring that food is safe. A plethora of news reports has exposed the inspection system's croneyism, lethargy and unfairness. More inspectors will not solve the problem. Consumers' freedom of choice, however, can.

In Virginia a human baby may be aborted at the mother's whim, but that same mother cannot go across the road and buy a glass of unpasteurized raw milk from the neighboring dairy farmer. The cry "freedom of choice" should ring from the lips as passionately for food, medicine and education alternatives as it does to destroy the unborn baby. If a consumer is willing to inspect the farm facility she should have the freedom to buy what the farmer has to sell. On-farm sales, at the very least, should be exempted from government inspection. I am not criticizing inspection for interstate transfer, but merely for on-farm sales.

Government inspection is primarily a matter of system, not quality. And the systems are engineered with "bigness" in mind. Many of the requirements are ludicrous or detrimental when applied to small-scale, on-farm processing. The critical issue here is not so much to eliminate large processors from inspection, but to free up the consumers' choice to buy and the farmers' ability to sell, within the parameters of small, on-farm processing with on-farm sales.

Our order blank contains a question: *"Do you want them dressed?"* Each customer must mark *"yes"* or *"no."*

The price is so much per pound, with no differentiation made between live and dressed birds. It's the same price. There is no law that precludes me from dressing chickens for you in my backyard if you bring them to me and I do not

137

charge for the service. Our processing is a gift, an amenity, that our customers ask for and for which we do not charge a penny. By not charging for processing, we do not fit under the legal definition of a processor. We are outside the code.

Customers are welcome to take the birds live, and that has actually happened a couple of times. This procedure keeps us strictly within the confines of the law and as far as we can tell keeps everything legal. It does mean, however, that if someone drives up to the house unannounced and wants to buy a chicken from the freezer, we do not sell him one. We're glad to take his order, however, and raise him one precontracted in the field. Consistency in this area is extremely important. We can, of course, give away processed chickens any time. Donations are acceptable.

Most states have an exception for on-farm processing and sales of poultry. In some states it is as high as 20,000 birds and in others it is as low as 1,000. This exemption is for processing the birds you produce on the farm where they are produced. It does not allow you to process birds produced by someone else. And it does not allow you to process the birds and then ship them off the farm. It does not allow you to cross state lines with the chicken. But for an operation like ours, it is an excellent exemption.

The exemption is certainly warranted on the basis of size. It is just a lot easier to keep a small, intermittently used processing facility clean than it is one that is used daily with thousands of chickens. It is only reasonable that some middle ground exist between the vertical integrators and a farm producing a few hundred birds for some neighbors each year.

Because our system does not fall within the parameters of processing facility definitions, we fall through the cracks of the regulatory process. A bureaucracy cannot regulate something it does not define.

The first thing some people do, before even raising the first chicken, is to run down to the health department and spill their guts. Look, bureaucrats are busy people. They essentially spend their time dealing with complaints. There is enough gray area here that at worst you would be asked to cease and desist, at which time you could find out just where you stand legally and then make those decisions. You are not going to be fined or jailed before plenty of preliminary give-and-take occurs.

I encourage you to just go ahead and start. Don't let your dreams be subject to some bureaucrat who has not a clue about consumer choice and good food. Don't get me wrong. There are some fine public servants out there, I'm sure. You just have to look real hard to find them. They tend to work on bigger projects. Chances are they won't even know you exist.

I have purposely remained somewhat ignorant about the inspection requirements so that if we ever get a letter from the bureaucrats we can just say we were ignorant. Like I said, there is enough gray area here that a little ignorance will put you in good stead. The fact is that many of these legal points are debatable. You can build quite a business and get some good loyal lawyers as customers before you even hit 1,000 birds. Don't try to hide anything. Stay open and up front about what you are doing because, depending on your state law, chances are you are perfectly legal as long as you market the birds live, do not charge for processing, and require customers to come to the farm to get the birds. The more customers you have, the more political clout you wield.

Our own experience attests to the validity of this position. Our farm has been featured in countless newspaper articles, magazine articles and television stories. I speak all over the state to groups containing legislators, health department personnel and the like. If anyone has gone public with an operation, it's us. But we have never been challenged in any way. We are careful to say that we sell only live birds by subscription sales.

Suzanne Ginger and Mike Evans in Farmington, IL were confronted with the toughest limitation in the U.S.: a 1,000 bird exemption limit. In just one year they received a two-year exemption for processing and introduced a bill in the Illinois legislature to raise the exemption limit to 6,000 birds. Much to their pleasure and surprise, they found many allies in the agriculture and consumer communities.

A law that frees up the consumer for more choice and the farmer for new markets has a tremendous positive sound. Who could be against that? The general argument against it is that someone may get bad chicken, and inspection protects the "general welfare" of the citizens. Of course, this does not address the fact that current production and processing models hurt people all the time. The main point you want to stress, though, is that you are not asking for all inspection to be lifted. All we want is an exemption for on-farm produced and processed birds, sold on the farm. If a customer comes to the farm and is satisfied with the quality of what he sees, he should have the choice to patronize that farm. To deny him that privilege is to assume that every citizen is an ignoramus, too stupid to evaluate cleanliness and integrity. That is a fairly unpopular political position to hold.

I am not an anarchist, and would caution you against foolhardy disregard for legal requirements. For example, dairies are heavily regulated. But even there, loopholes exist. For example, a lady in West Virginia markets her milk as pet milk, and customers sign waivers acknowledging that the milk is "not for human consumption." She's been doing this openly for one year and says it's completely legal.

Whether it is or not, I don't know. I'm not an attorney - and even attorneys disagree over many points of law. Probably if this lady sold milk across state lines, she would be vulnerable. Size really is an issue.

While I don't advocate running roughshod over clear legislation, neither do I advocate tucking tail and running from every obstacle or every threat. Life is full of risks, and we must be courageous enough to take some risks to achieve success.

Chris Wieck in Umbarger, TX, decided to build a small processing plant under PL 90-492, a federal processing law that allows up to 20,000 birds per year to be processed in a state inspected facility without any inspector being present during processing. He hasn't had any problem with inspectors, although he encouraged people who wanted to go this route to work with authorities on blueprints until everything meets government approval.

This state inspected facility allows him to sell birds within the state, both retail and wholesale. The minimum cost is about $50,000, but "it's a much better investment than a John Deere tractor," he quipped. He figures it will have a four-year payback. He is doing as much custom processing as he is his own. "We get customers from 400-500 miles away for custom processing. We could charge anything we wanted to," he said, although his price is $1.25 per bird.

He hires some local high schoolers and three women to help with the processing. He processes every Saturday morning year round and has now added an on-farm store. Amarillo is 30 miles away and Lubbock is 100 miles. Those provide his customer base, and folks come to the farm to purchase the birds. He added a used walk-in cooler and sells most birds frozen. His wife, Sara, who handles the farm store, reports that it is "going great. Folks are hungry for some real food." They correspond with customers via newsletters and have added their beef, pork and eggs to the store fare.

Chris reported that retail farm sales puts a "fire in your soul. The personal touch is what I enjoy and it makes you feel good. It's a side of farming I didn't know existed." Amen, Chris.

Let's be aggressive about taking control of our business and our lives. This is not the time for timid people to sit on their hands. Just be uncompromising about doing a good job, being honest, and building loyal customers, and I doubt that you will have any trouble. Renew that old frontier spirit and go for it.

PROBLEMS
(IT'S NOT ALL ROSES)

Chapter 18

The Learning Curve

Now that we have been successful at raising pastured and home-processed poultry, many people are interested in duplicating our model and we encourage that. It's the ultimate human praise. But I always smile to myself at how many people think it sounds so easy. Lest we appear to have just stumbled into a neat agricultural scheme, allow me to share some of those major learning curve experiences that accompany pioneer efforts.

The problem the first year was processing. We started with a big kettle over a wood fire. As a teenager, I had raised layers in these portable pens and dressed the culls. I cooked the birds, picked off the meat, and sold it as precooked, boneless chicken at a local farmers' market. A real delicacy, it sold easily to convenience-minded consumers. At 14, I had purchased an old galvanized work table and a tiny table top picker. That was what I had used to pick and eviscerate cull laying chickens. By selling just cooked, boneless meat, I didn't have to worry about a few feathers staying on the skin. The skin pulled off real easily after cooking.

But these broilers were a different story. We didn't know about the 145°F water, and with the wood fire varying the water temperature, it would burn one chicken and not scald the next one enough. It took Dad, Mom, Teresa and me to process about 10 birds an hour, mainly because of feather picking problems. That was when we decided we were going to have to do better on the processing.

The following year, therefore, we raised fewer birds (300) and rented the local small locker plant's poultry processing room. But that year, the birds were no larger than pigeons. They just wouldn't grow. We concluded that we had crowded them too much in the pens. We put about 150 in a pen, instead of 100 or fewer. That was when we learned about crowding stress. We transported the chickens to the locker plant in our dump truck. When we arrived, 16 of the biggest ones were dead. They had piled up against the tailgate and suffocated. We learned the importance of crating these birds so they don't pile up during transport.

Processing went well, though, despite the small birds. But the hassle of taking scales to weigh the birds, being away from the farm, and meeting customers there made us conclude that we needed to process as efficiently as the locker plant, but do it at home. We sold some equipment and invested in a brand new automatic feather picker.

By putting an ad in the rural electric cooperative magazine, we located a small thermostatically controlled scalding tank on the other side of the state. With our new picker and a 2500 watt scalder, we were ready for year number three and let the orders go up to 800.

As the season got underway, the weather

turned terribly windy and cold a day or two following the arrival of our first chicks. A day or two into this last blast of winter, chicks started dying. They would just drop over and I was desperate to find the cause. It was pneumonia, due to cold, drafty conditions inside the brooder house. As soon as we got a warm, sunny day, we moved the remaining chicks out into a little pen in the yard. Made of some scrap boards propped up with bricks, the enclosure had no roof on it, affording the chicks full sunlight luxury. Meanwhile, I tightened up the brooder house and purchased more heat lamps.

Things went fairly smoothly after that until the following year when I went out to the brooder house one morning and noticed that the chicks, huddled under the brooders, were extremely quiet. Chicks should be boisterous, chirping and singing as they go about their work. I couldn't tell that any were missing but they certainly acted scared. The temperature was fine.

Later that day, we found a pile of dead and dying chicks under some old feeders and gunny sacks in an adjacent room where we kept barrels of feed. After consulting with several neighborhood farmers, we concluded that it must be rats.

That night, we moved an old army cot into the feed room, cleaned out all the junk, and prepared to do battle with the rats. I took up residence in the chicken house, sleeping there every night for a week, with a broken baseball bat as my weapon. One night about midnight something startled me and I rolled over to see a rat about a foot long climbing up the outside of the brooder house wall. His white stomach was about a foot from my nose. But he didn't bother the chicks.

That year I plugged every crack and every

hole I could find to make the brooder house rat proof. A couple of years later we had a similar attack, up through a gnawed hole in the floor, through the deep bedding. Rats will take chicks and just stuff them down a hole, many of them still alive. After surviving the rats, we wondered what else could go wrong.

We were soon to find out, for the next catastrophe occurred just after I had moved the first batch out in the spring. A 2-inch rain that night turned depressions into ponds and water was running everywhere. I pulled out bucketfuls of dead 3-week-old chicks. That was when we learned about the need to stay on top of things when earnest rain sets in for a long period and water begins running across the top of the ground. Over the years, we have gone out at numerous times in the middle of the night, Teresa carrying a flashlight and me carrying two bales of hay, to put down a dry hay pad and gingerly scoop wet chicks up onto the dry hay. Sopping around inside that cage, with the manure and panicked chicks, is certainly not fun. But that is the price for chicken worth eating. There is no free lunch. Fortunately, it doesn't happen often.

During the summer of 1987, we had a terribly hot spell. I went out in the morning to move the pens, and found several dead birds at the back of each pen. We were just a few days from slaughter and these were the biggest birds. They had suffocated from the intense heat the previous day. By propping up the rear of the pens during the early afternoon, we solved that problem. (Photos 18-1 and 18-2)

During these early years we were changing the feed ration, trying to eliminate all syn-thetic vitamins and minerals. We used kelp and switched the straight soybean meal to part soy,

Photo 18-1. *In extreme heat, we pick up the back end of the pen and prop it up to let air in to the birds.*

Photo 18-2. *For ventilation, the crack can be quite large before mature birds can get out. The prop assembly allows infinite height adjustment.*

148

part peanut meal, and part roasted soy.

Toward the end of 1988, when we were three batches away from the end of the season, the little 12-day-old chicks began showing signs of some sort of paralysis. Their curled toes put them off balance and they would rock back on their hocks. In advanced stages, their legs would trail out behind them and they would scoot about pathetically on their breasts, flapping their wings against the litter. More were coming down with this strange malady every day.

Whatever the problem was, I figured getting them outside would cure it. Even though the weather was threatening, we moved them outside, and promptly had a deluge that left several dead of suffocation. But the paralysis continued until it began slacking off around day 25 and seemed to run its course by day 28. We lost about 30 percent of the chicks. What could be the problem?

The hatchery had warned me that these chicks were from a new breeder flock. Pullets, because they lay smaller eggs, produce smaller chicks. The hatchery warned me to treat these chicks like "preemies." They were smaller, but otherwise appeared to be healthy. Then this malady occurred.

The next batch did the same thing at day 12. I began noticing some birds going down. After consulting my *Merck Veterinary Manual*, I determined that it was curly toe syndrome caused by a riboflavin deficiency. Grabbing some beef liver from the freezer, I fed it to the chicks and the problem stopped immediately. Even the paralyzed chicks began walking again. Since putting brewers' yeast in the ration, we haven't had the problem.

In 1989, everything was going fine when all of a sudden the chicks in the brooder house quit eating feed and began drinking water like fiends. They absolutely went crazy over water. They jumped in it, got soaking wet, then piled up and suffocated trying to get warm. They wouldn't eat. I was desperate. What could this new problem be? I noticed the feed seemed a little discolored and heavy. The bottom of the feed tank held what came out of the truck's auger first. Could there have been a residue of something toxic in the auger? Normally the feed trucks just hauled dairy pellets, which were fairly compatible with the broiler ration. The bigger birds out in the field, on the same batch of feed, were fine. I brought in some feed from the field and the little chicks immediately began settling down. They ate it like they were starving to death. I called the feed mill.

The truck driver, prior to delivering my feed, had made a haul of mineral supplement to a local dairy and had not cleaned out the auger between that haul and mine. The residue left in the auger came out first in my tank, and as I got to the bottom of the tank, comprised a major portion of the feed I was giving the chicks. They were being fed calcium and salt. No wonder they were thirsty! We lost about half the batch. Needless to say, the truck drivers are very careful to follow clean-out procedures before they bring chicken feed to us.

Another point in our learning curve occurred after a customer called with car trouble and we had to refrigerate chickens overnight. We learned that the birds must always be put on wire racks. When placed on the solid bottom of the refrigerator, or on a solid shelf, air movement is restricted underneath the birds and it will spoil them even overnight. Poultry is extremely perishable, and must be protected from even a few hours without adequate ventilated refrigeration or ice.

What else could go wrong? In April of 1990 we had had our first batch of 1200 chicks a week when we received a wet, heavy, 10 inch snow. April snows are highly unusual here, but this one came anyway. Around midnight, the electricity went off. The temperature was right at freezing. The chicks were just a few days old, and suddenly were plunged into darkness and no heat. We grabbed the shipping boxes the chicks had come in and madly began scooping chicks into the sections, trying to stay ahead of the birds as they panicked and piled up inside the brooder house. In an hour, we had all the chicks reboxed and stacked in the house, near the fireplace.

I slept on the sofa all night to keep the fire going. All the ruckus kept me from sleeping too soundly. By morning, the power was still out and we had 1200 thirsty, hungry chicks in the house. What to do?

There was no alternative but to turn the kitchen/dining area of the house into a brooder facility. I went to the shed to get a roll of poultry netting which I planned to tie with baler twine to the legs of the furniture, creating a big circle. Newspapers would be okay for litter until the power returned.

When I came back in with the netting and walked in the door, the lights flickered and came back on. Teresa shouted "Hallelujah!" We danced a jig, donned coat and boots, and began taking out boxes as fast as we could. In a few minutes, the chicks were all back in the brooders, as happy as could be. The total loss was about 20 birds.

What else can go wrong? Who knows? The reason for sharing this is to make you aware that ours has not been an enterprise built of ease, but one fraught with a host of Waterloos. Each one taught a valuable lesson. We don't want them to occur again, and by God's grace they won't. Be assured that other crises await us of which we know not. Won't it be fun to see what they will be and what we will learn from them?

Chapter 19

Sickness and Disease

Vaccines, antibiotics and hormones, housing sanitizers and chemicals so permeate modern poultry production that the idea of raising healthy birds without these crutches sounds incredible to most poultry producers. The risk of sick birds, they say, far outweighs the advantage of nontoxic production. But in all our years of producing poultry, we never had one outbreak of disease or sickness. The conventional counterparts in our area have had several.

That is not to say we never have sick birds. Anytime we deal with living things, death is inevitable. But this mortality accounts for only 2 or 3 percent of our total. Most of our losses, when they occur, are due to weather or predators. And yet, even with these figured in, our mortality is no higher than industry standards, and is normally lower. Anything under 10 percent is considered acceptable.

The first concern is toward sterilization of the house between batches. When we clean the brooder house, we only use a scoop shovel. We do not sweep out the corners or disinfect it in any way. Beyond what the naked eye can see is a tremendous battleground. The good critters are fighting the bad critters. Given a healthy environment, which includes a carbon percentage high enough to eliminate all traces of ammonia smell in the litter, the good critters will hold the bad ones in balance.

Sterilizing agents are broad spectrum and kill off good bugs as well as bad ones. Environmentally harmful substances hurt the good bugs more than they hurt bad bugs, and the surviving bad bugs develop resistance to the spray, fumigant, disinfectant material and become more virulent. This resistance factor makes what scientists call "super bugs," which are much harder for the good bugs to hold in check. They are also harder to kill with poisons. Sterilizing, then, can easily set the stage for epidemics.

When a new batch of chickens arrives at the brooder house, they will stay healthier if there is a good population of good critters in place to maintain balance.

Vaccinations are unnecessary if the production model is healthy enough. Animals are born with immunities, and the more we tamper with those, the more we reduce the natural immunities and the more likely we are to develop a generation that is immunodeficient. Does that sound familiar? Today, with all the vaccination going on in the livestock industry, it is no wonder that the human population is becoming more and more immunodeficient. The reason the animals need vaccination to survive is because the production systems are not meeting the needs of the animal. Return the animal to a natural, healthful element, and vaccinations are neither needed nor desirable.

Perhaps the single biggest sickness problem

is pneumonia in new chicks. Temperature fluctuations must be minimal on these new arrivals, and especially drafts must be eliminated. Putting burlap around the hover and leaving just a couple of access holes for the chicks to come out and eat and drink can help cut down on drafts where they are sleeping.

They just need a completely draft-free place to sleep and lounge that is within a 10-degree tolerance (85°-95° F) all the time. Greater fluctuations in temperature will bring on pneumonia. The telltale signs are light birds, discolored, spindly-looking legs, and a hunkered-down appearance.

The chicks should feel solid when you pick them up. You want some bulk and weight. A chick with pneumonia will feel like a cotton ball; there is just no weight. They feel like they have wasted away. A healthy chick will walk upright, standing tall. A chick with pneumonia will appear stunted; its wings stick out a little, and it has a humpback, or hunched over, appearance. The definitive sign is the legs, which should be deep yellow on a healthy chick, but turn brownish orange on a chick with pneumonia.

Short of antibiotics, I do not know of a cure for birds with pneumonia. I think everyone who raises birds needs to experience it one time just to realize how seriously to take the admonition: no drafts and a warm, dry place. It's another one of those situations where until you've experienced it, you simply cannot understand the importance of this admonition. If you see two or three birds showing pneumonia, realize that you are to blame, not the chickens. Make sure their temperature is not fluctuating drastically and get rid of all drafts. They don't mind coming out in the cold - even 30° F to eat and drink - as long as it is not drafty and they have a warm draft-free lounge area.

Generally the pneumonia will run its course in a week. That's the tragedy of it. By the time you realize they have it, it's too late to save many of the birds, and you just stand and watch helplessly as more birds show the telltale symptoms and die within 24 hours.

Take heart, though, because seldom will you lose more than 50 percent, even in a bad outbreak. Take your lumps, learn your lesson, and do better next time.

Once the birds are a week old their tolerances seem to take a quantum leap. They still should not have drafts, but they can take 20° or 30°F variations, they need not be warmed beyond 75°F, and they are just about over the hump from ever getting pneumonia.

The only other time you may see pneumonia is out in the field during long bouts of extreme damp, cold weather or when dramatic fluctuations (more than 40°F) occur between day and night temperatures. The symptoms are the same. I call it "shrivel up syndrome." That is exactly what the birds appear to do - just shrivel up and waste away.

If you cut them open, you will find their entire body cavity full of fluid. Often they smell funny, too. The probiotics are invaluable in preventing pneumonia.

Coccidiosis is a major concern of the poultry industry. Coccidiostats in feed or water are thought today to be as important as the nourishment itself But this disease, called cocci for short, is caused by unsanitary condi-tions, primarily attributable to fecal contamination. In a typical chicken house, fecal dust is everywhere. In fact, the house air has a distinct "hazy" appearance, and most of these suspended dust particles are pieces of manure. The waterers become coated, every minute of the day, with more fecal dust, which goes into the water pan or onto the water nipple. Feeders receive a constant sprinkling of fecal dust.

And worst of all, the birds ingest huge amounts of fecal dust in the air they breathe. It coats their lungs and their nasal passages. It gets into the digestive tract via food and water. The bird is full of fecal contamination inside. The dust coats the bird on the outside, in its eyes, on its beak, in its feathers and clear onto its skin. Is it any wonder that housed birds normally require coccidiostats to survive? In our pens, there is absolutely no dust, zero fecal contamination except for an occasional stray manure dropping on a feeder, which is completely different from fecal dust. I can easily pull it out at feeder filling time, and meantime, the chickens avoid it.

Antibiotics do not build body tissue. They simply mask symptoms of deficiencies. Many diseases are caused by inadequate nutrition, as well as by improper production systems. The symptoms - generally some type of infection - can be masked by antibiotics. The antibiotics do not remedy the problem. They simply allow the bird to survive until slaughter. The bird may appear healthy, but it may be quite deficient. In fact, if the antibiotics are removed from the typical factory poultry house, the birds will begin dying like flies.

These antibiotics produce resistant pathogenic organisms and pass that resistance on to humans. Many studies have confirmed that

157

Americans actually ingest more drugs through the meat they eat than from the pills and injections they take. We have chosen to fight the diseases at the sources, and so far have found that every single one will respond to nutritional or production modifications.

All of our efforts notwithstanding, we do have sick chickens. Some batches have practically none, while others will have several. The reason we have fluctuations between groups is because we do not mask deficiencies with drugs. Variables in the breeding hen phase, over which we have no control, affect the disease propensity of chicks. For example, one breeder hen house may have its feed tank in the shade, while another may have its bulk tank in the glaring sunlight. Synthetic vitamin packs lose potency fast when subjected to heat, and the poultry industry can document cases of chick deficiencies after a severe hot spell. The deficiency passes right through the egg. Then the chick will be born with vitamin deficiencies.

Breeder hen houses can vary with regard to air flow, affecting fecal contamination; sunlight, affecting pituitary function and vitamin metabolism; parent material of feed; drinking water quality; and a host of other things. While most growers mask these variables with antibiotics, we do not.

For sick birds, we use a hospital pen. (Photo 19-1) This is a separate, small pen that we keep near the yard by the house. It is just like the bigger field pens, except on a small scale. We can quarantine the sick birds and give them special attention. It is amazing how they respond.

Their area is small, allowing them proximity

Photo 19-1. *The hospital pen.*

to feed, water and fresh pasture. We can keep an eye on them easily and make adjustments. Most of them respond dramatically in just 48 hours. In fact, most of them can eventually go back out in the field pens and perform fine. They just need a little extra attention. Perhaps they could be called late bloomers, or slow learners. We call them our LD (learning disabled) birds. After all, it is not politically correct to accuse them of being sick.

If you have the house space, you may want to use deep litter and a stationary house for a hospital pen, or even a special corner of the brooder house if it is big enough to accommodate the birds. These birds normally have leg problems and do not move well. With our new larger brooder facility, we now use our two 80 square foot brooder huts for hospital pens. This gives us one per batch and we don't have to move sick birds on pasture. They heal equally well.

We make sure these birds get grass clippings, beef liver, occasional handfuls of weeds
and grass with attached soil from the garden. A
key part of the healing is to keep taking out the
birds that get well to reduce the competitive
pressure on the ones who remain sick. That
constant reduction in competitive pressure keeps
the slower ones making progress.

Some stay in the hospital pen until slaughter. A few may be good enough to sell, but many
of them end up on our table.

Common problems for which birds are culled
at commercial processing plants - breast blis-
ters, skin abscesses and abrasions, tumors and
the like - are virtually nonexistent in our
broilers. Again, these maladies indicate a
production model gone awry, and will vanish as
that model more exactly meets the needs of the
bird.

In the final analysis, I do not claim to be
an expert on poultry diseases. We have had the
unparalleled joy of simply not having to deal much
with that aspect of the business. We do not
believe that is by accident. We believe it is a
direct result of a healthy production model.

Chapter 20

Predators

Perhaps no other issue concerns would-be poultry graziers more than predation. Anyone who has lost sheep to dogs and coyotes is understandably concerned about what a predator would do to a pen of 100 succulent 3-week-old broilers out in the middle of a field half a mile from civilization.

Many who believe all chickens should be free ranged picture a pastoral setting of blue sky, 75 degrees, a few puffy clouds dotting the horizon, green grass carpeting the hillsides with cows contentedly chewing their cuds under an apple-laden tree. The chickens scratch unhurriedly in the pasture, clucking their contentment while the farm dog, lying prone on the farmhouse porch, moves only to snap at a pesky fly buzzing around his nose. These folks have been patronizing Walt Disney too much.

This setting occurs only in the mind. In reality the landscape pulses with the power of successional plant communities, the clouds billow up into black thunderheads, and predators lurk at the field edges during the day and roam the fields at night seeking tasty morsels to satisfy their carnivorous cravings. Anyone who has seen what rats can do to baby chicks, what opossums can do to older chicks, and what weasels can do to mature chickens, realizes that prudent protective penning of animals is as beneficial and sensible as locking the door of our house when a burglar is in the neighborhood.

On our farm, we do everything possible to feed and maintain a happy, healthy predator population. I have no fear or dislike of predators. They fulfill a desperately needed function. They eat field mice, rabbits and all sorts of things. They act as a natural check on animals that otherwise would overmultiply. By not using pesticides, herbicides or artificial fertilizers that harm the critters that inhabit stream, field and wood, we encourage a healthy population of predator food. Cattle are penned out of the forest, thereby rendering it fit habitat for all the different wildlife species.

We do not hunt rabbits, but encourage brushy edges around the fields. Years of pond building have ensured plenty of water for aquatic species. Raccoons have a place to wash their food. Rational grazing of livestock produces a dense pasture sward conducive to mouse and vole populations. Stockpiling winter forage provides warm habitat for these pasture critters to feed predators even in the winter time. The point of all this is that we have tried to maintain a balance and do all we can to have a satisfied predator population.

But there is a rogue in every bunch. If a predator cannot be satisfied with the food we've tried to give it, then we deal with that particular wayward individual. That, in short, is our philosophy regarding predators. One sidelight: it is a shame that the predator's predator, man, has been all but pulled out of the natural balance by the animal rights extremists. As a result, predator overpopulation is becoming more of a problem, as is rabies. It all has to work together, each part of nature filling its vital role. Animals are not human.

Prevention, of course, is the best route. Predators tend not to walk across the open field. By mowing or grazing before and behind the chicken pens, we maintain a short sod and thereby discourage predators from even traversing the area. A great horned owl will actually snatch a weasel right off the ground in the open field.

By far and away the predation is more likely to occur if we have the pens right up against the woods at the end of the field. We carry small boards on the pens that we call plugs. An assortment of lengths allows us to plug depressions in the ground that exceed one inch between the ground and the bottom of the pen. (Photo 20-1.) In all the years we've been pasturing poultry, only once or twice has a predator ever pulled away a plug. And they don't try to get in through the top.

Generally they try to get at the chickens through the mesh on the east end, where they can look in and see the bids sleeping. The critical period is the first few days when the chicks are small. When they sleep, the birds on the outside of the pile get pressed up against the wire mesh as the group snuggles together to stay warm. A predator can grab a head, wing or leg through the mesh. Once the birds get bigger, a predator can't inflict serious injury through such a small hole. Injury does not always mean death, either. We've had chicks scalped or lose wings up to the second joint heal up and grow out fine. The other birds pecking at them is simply not a problem. Our birds

Photo 20-1. *Plugs under the pen on uneven ground keep predators out and the birds in.*

are not cannibals.

If we do get an attack like this, which accounts for 90 percent of all our attacks, we set leg traps the following evening. Predators always attack the same pen on the second night. It is not necessary, therefore, to set traps around all the pens. We try not to disturb the attacked pen, even leaving any dead chicks lying where they were left. We literally mine the pen with half a dozen or more leg traps, set on the ground up against the pen, especially adjacent to the area of previous attack, and lightly covered with some grass to camouflage the steel.

The traps tend to get rusty and where the steel has been banged it will be shiny, reflecting moonlight and alerting predators that something is awry. Both problems can be controlled by boiling the traps in walnut hull or sumac water.

Using a three-gallon kettle, we put in a quart of walnut hulls or sumac seed heads, then fill the container with water and bring it to a boil on the stove.

To maintain peace in your household, take the kettle outside before putting the traps in. You don't want to put the traps in boiling water because the heat will take the temper off the metal. But you do want the water to be just under boiling. Submerge the traps and let them soak until the water cools down. Take them out with a stick, rubber gloves or tongs. The heat gets rid of the rust, the hulls or sumac gets rid of smells, and the color darkens the traps so they won't glint in the moonlight. You want a freely moving trap that will activate at the slightest weight.

Do not touch the traps with your bare hands. Most predators, especially foxes, have a keen sense of smell. These animals are smart, believe me. I store wooden pegs to hold down the traps, the traps themselves, and my neoprene working gloves in a bucket. Leather gloves hold odors and should not be used. When you put on the neoprene or rubber gloves, do not touch the fingers with your bare fingers, but work the gloves on carefully like a surgeon would.

We have never had much success using live animal traps. Each predator has its own attack characteristics. The way you respond to the predator depends on what the predator is. I can only describe the predators we have experienced. To my knowledge, we have not had a coyote, bear or dogs. But we have had a wide range of experience with others.

Rats are only a problem when the birds are tiny. You will generally notice a couple of chicks with telltale bloody scratches or tears. They are such small wounds, most of the time you will just dismiss it as a superficial injury. When you go into the brooder in the morning, you may notice that the chicks are abnormally quiet, almost hushed. They may not be out on the floor feeding as actively as normal, but rather cowering under the hover or over in a corner.

Often rats come up through the floor, but certainly they are not particular about where they gain access. They will usually take two or three chicks a night, unless it is a terrible rat infestation that can take up to a hundred in a night. They generally take more than they can eat, and stuff the chicks down their hole, alive, for later consumption. You can hear a quiet chirping of the chicks, perhaps behind a wall panel or coming from under the floor.

One way to check for rats is to go out an hour or so after dark, real quietly, and sneak up on the brooder. If you see rats inside, you know you have problems. They will continue to take chicks even up through the fourth week. Often you don't realize you have a problem until you take out chicks and find you have far fewer than you should have.

The best cure is to plug up the holes. Quarter-inch hardware cloth works well around eaves or windows. Anything wooden can be gnawed through, but that takes a long time. Pressure treated lumber helps. You can set traps on ledges up away from the chicks.

Opossums and raccoons are by far our two worst predators. Opossums, which of course we call 'possums, are spiteful. They are not nearly as interested in eating the birds as they are in making a mess. Our worst predator attack ever was from a 'possum shortly after we put some birds out onto pasture. The ground was terribly uneven and we had a 6-inch space between the bottom of the pen and the ground on one corner. We put boards against it, but a 'possum knocked them down and in scarcely half an hour killed all 70 birds inside. We had gone to town on an errand, and when we returned, about an hour after dark, I went out to check the birds and found the 'possum looking for a way to get into another pen. Needless to say, I made short work of that dude.

He had strewn carcasses for 20 feet around the pen, many with their intestines all strung out away from the carcass. It was dreadful. 'Possums seem to just enjoy killing for pleasure. They prefer to eat the innards rather than the meat, and will even kill adult birds.

A raccoon, on the other hand, does not kill for fun. He generally eats the carcass on site, and only kills what he wants to eat. Foxes prefer to carry the carcass away, and only leave a pile of feathers on site where the kill was made. But we'll get to foxes in a minute. Raccoons are extremely strong, and can yank at poultry netting like a person. If it is not stapled down well, they can jerk it loose and get into the pen. We have never had a raccoon actually get into the pen except once, when he was already in a trap. Usually they will dig enough to get a paw under and grab a leg or wing, eat it off, and leave the rest of the carcass inside the pen.

If they can get the bird outside, they will eat what they want - always eating the head first - and leave the rest. The next night, they will not finish off the old carcass, but kill another fresh bird. Headless birds indicate raccoon kills. Birds with heads on but guts strewn around indicate 'possum kills. Raccoons are strong enough to sever major thigh bones with their teeth. They can be mean rascals.

A smart old raccoon can be a tough animal to catch, but if you are careful setting your leg trap, you will get him. 'Possums are easy. You can trap them or walk up on them without any trouble.

A fox generally will not come around the pens because there is just too much human scent around. Foxes are wary creatures. I think that is one reason why we have not had coyotes in the chickens. Foxes like to carry the bird away if they can. They will also dig, and in fact seem to be the only animal smart enough to figure out that the way to get the birds in the back of the pen is not through the front poultry netting, but rather by digging under the aluminum sides adjacent to where the birds are sleeping.

A fox that is getting into chickens, fortunately, becomes lazy about hunting other food. He quickly establishes a routine and comes back repeatedly. Any time an animal establishes a routine, he's much easier to catch.

In all our years raising chickens, we've only had one weasel, which pound for pound is supposed to be the most vicious mammal in the world. A weasel does not want to eat meat; all he wants is blood. If you see dead chickens but can't readily see any wounds on the birds, examine the neck closely. You will see two holes in the neck, right in the jugular vein. That is a dead giveaway (no pun intended) for a weasel. One weasel attack, which occurred about 5 years ago, happened a few days before slaughter. He got inside the pen and killed six big birds in one night. But I set traps and got him at the same pen the following night. He was already in a couple of traps within an hour after dark.

Great horned owls can be a problem when the birds are tiny. I've only had to kill one. They are only a problem when the birds are tiny. A scalped chicken, still alive in the morning, or a dead carcass missing only a wing, indicates an owl attack. Again, a leg trap is very effective.

Snakes have never been a problem, although I know they will kill chicks if they can get into the brooder.

In all our years raising chickens, we've only had one fox. He outsmarted all our efforts to trap him. He didn't kill a large number of birds, but consistently killed a couple or more per week. He would reach under the bottom board and grab a leg, pulling it off. In the morning, I would find the dead chicken in the pen, missing a leg. He never could get in the pen, even when cracks were fairly large and he tried to dig. It was extremely frustrating not to be able to stop him.

I tried sleeping in the pens, with shotgun ready. I tried sleeping in the pickup cab, with gun and headlights ready. I slept in the feed tanks and listened to him bark at me from the edge of the woods. He always knew I was there. I walked up at varying hours of the night, trying to surprise him. I had a professional fox trapper set traps in the field, at the edge of the pens, baited and unbaited. Nothing worked. We tried a recorded rabbit caller to lure him in. Still nothing.

Finally I invested in a powerful handheld rechargeable spotlight. What a great tool! From 300 yards you can illuminate an area like day, and any eyes in that bright light stand out like two huge lightbulbs. As today's teenagers would say, it is awesome. The light disorients the animal. I was able to walk up on him and stop the problem with hot lead. The 500,000 candlepower light cost well under $100 and was worth every penny.

In the course of all this, we added a 6-inch high fringe of quarter-inch hardware cloth around the northeast quadrants of the pens, to keep anything from reaching through the poultry netting when the chicks are small. I think it is a worthwhile addition to the pens if you have a serious predator problem. Generally the birds tend to sleep in that quadrant when they are small.

Beyond that, predator proof electric fencing can be constructed with 8-strand high tensile wire and a powerful charger. The wire should alternate between cold and hot, with the cold wires grounded direct to the ground pole of the charger. The lowest wire is a ground. Most predators want to climb, and by touching a fence hooked up this way, it is equivalent to putting your two hands directly on the two output poles of the charger. For 20 acres, which is the area required for 10,000 broilers, the investment for such a fence would be $800-$2000, depending on how much of the work and/or posts you supplied yourself. If predation is a real problem, this is a small price to pay for total peace of mind, and an option we will pursue if we continue to have problems, which we have not. Such a fence will keep out 'possums, raccoons, foxes, coyotes and anything else that may cause problems except owls, and the mesh on front stops their attacks.

One other trick I've been told is to use a treble fishhook. Put one of the hooks in a dead bird and leave it as bait. The predator hooks the other hooks in his mouth and that eliminates the problem. Sometimes the predator will actually swallow the hooks as it gulps down the meat. We tried this for the fox, but he ate around the hooks.

It is quite possible that predators will become an increasing problem as the trapping industry wanes in the wake of Walt Disney and the humanizing of animals. The market for genuine animal pelts is nearly gone, and this is unfortunate. It's another case of overreaction on the part of many because of the abuses. We farmers and hunters are often our own worst enemy.

We want to stay in harmony with the predators and do everything possible to ensure them a healthy wild food supply. But when a rogue steps over the line into the domestic scene, he must be dealt with like a hardened criminal. To do less is folly.

It's wise to just go out once in awhile with a gun and the spotlight, just to make sure things are quiet. I have come up on raccoons and 'possums that way. Generally, if anything is going to come around, it will be there within two hours after dark.

We've gone entire years without a single predator attack. It is important to note that this is simply not a major problem. It does happen, that is true, but it is not something to be worried about. We stay prepared for it, and do all we can to prevent it, but it is a rare morning indeed when we lose a bird to a predator. And we run these pens nearly a mile away from the house at times, way beyond sight of any houses. We are not just running the pens in the backyard where we can keep a close eye on them.

The safety and security provided by well

constructed pens cannot be overemphasized. We
can only guess at the kind of scratching and
pawing that goes on out there some nights. But
the fact is that seldom is a predator successful
the first time, and virtually never the second
time. He doesn't get that chance.

Chapter 21

Weather

Probably the single biggest reason for confining broilers in stationary houses is to reduce the weather variable incumbent in any kind of pasturing setup. There is no free lunch. The down side of producing better chicken is a higher degree of management, including dealing with the vagaries of the weather.

Dad, who flew airplanes in the Navy during World War II, frequently spoke of a flying instructor who told poor performing student fliers that "you've got to do whatever it takes to make the plane do what you want it to do." Don't complain about the airplane; make it do whatever is necessary to get the proper response.

It's much the same way with pastured poultry and weather. You just have to get out there and do whatever it takes to keep the birds warm and dry. It may be messy, wet and cold. It may be midnight. But that is the price we poultry graziers pay for producing a broiler that is fit to eat.

The biggest single problem is rain. Although the pens are three-quarter covered, that doesn't mean that water can't come in underneath or that the birds will always get under the covering. After taking the little chicks out of the brooder house to the pens, it is wise to check on them about dark for the next couple of nights and make sure they are sleeping under the roofed portions. A couple nights of gentle repositioning is all it takes to establish a pattern for them to sleep under the roofed portions. Usually they want to sleep around the feeder anyway because that is where they are when it begins to get dark. They tend to stay in the place that stays lightest longest, and only a few of them will seek out the shelter of the totally enclosed back of the pen.

If the young birds get soaking wet, they get cold, no matter how warm the temperature. Few things are more ugly or vulnerable than wet chickens. Dry feathers are their insulation, and when that goes, not only do they have no insulation, but the dampness extracts heat from the body as it evaporates. When they get cold, they snuggle up together to get warm. As they snuggle together, the ones on the outside keep pressuring the ones on the inside until they begin piling up and suffocating. The big losses are not directly from being wet or cold, but from suffocation as the group piles up to get warm.

As the birds mature, toward 7 weeks old, they will not pile up when they get wet. They tend to just shut down. They hunker down, don't eat or drink, and slowly freeze to death. With the high metabolism on these birds, even 12 hours without eating can be catastrophic. They must stay active. Of course, the bigger birds can take quite a bit more abuse than the younger ones.

A quick half-day thunderstorm hardly ever causes a problem. The problem comes when a cold rain sets in for several days. And the worst thing that can happen is for the ground to become so saturated that runoff comes into and through the pen so that the birds slog around in a shallow pond. That requires drastic action, immediately.

The best solution is to take dry hay, climb inside the pen, and spread out the hay, fluffing it up, and then gently push the chickens up on the hay. It is not necessary to cover the entire area under the roofing. But hay should be placed around the feed trough and enough toward the back of the pen that all the birds can climb up on it.

That is one thing the birds do have enough brains to figure out. It is amazing how soon they will all be up on the hay pad, preening themselves and drying out. The steam will rise off the group as their body heat helps their feathers dry out. And although their feathers may be dirty from mud, once dry they will insulate and the birds will be happy and comfortable.

The main goal is simply to give the chickens a pad big enough to let them all break ground contact. They simply cannot slog around in water and live. They will die as fast and as sure as if they were slaughtered. It is amazing how resilient they are when they can break ground contact. If hypothermia has set in so that a bird is alive but totally stiff, normally it can be revived by bringing it back to the brooder and placing it under heat for an hour.

If the rain comes from the northeast or southeast (against the poultry netting) extra procedures may be required, especially if there is a breeze. Against the open portions of the pen, the draft can be terrible. A few pieces of old scrap metal roofing or two-foot strips of plywood placed against the pen side will keep out the wind and allow the birds' body heat to fill the pen. Fortunately, these procedures need not be taken frequently, but when they need to be done, there is no substitute or delay. Those birds must be kept warm and dry.

If the weather is cold and dry, the birds can take freezing or below freezing temperatures at three weeks. If it gets below 20° F, however, even mature birds will begin suffocating in their efforts to stay warm

The other extreme - heat - rarely causes problems unless a sultry day gets up over 90° F and the birds are more than 7 weeks old. Small birds can take nearly any heat. But as the birds get older, heat stress can cause acute reactions and can result in death. If the day will be a heat killer, the best cure is to go out at midday and prop up the enclosed end of the pen as high as possible with a couple of boards without letting the chickens out. This crack, especially if the grass around the pen is short, works with the reflective aluminum top and sucks cool air inside by convection, creating a strong, cool breeze. This procedure is quite foolproof.

The birds adapt well to gradual fluctuations in temperature or moisture. They seem to get acclimated to trends. In other words, if we have a heat wave that started when the chicks were small, so that they have been used to being warm, they can take far more heat when they are large. But if it has been in the 70s and 80s for several weeks, then suddenly spikes to 95° F one day when they are more than 7 weeks old, they will need special attention. Plan to raise that pen to give them more air. The same is true with cold. If they've been used to being a little chilly, they can take some low temperatures. But a warm spell followed by a sudden high pressure drop can cause problems, especially when the birds are less than 5 weeks old.

Be aware of huge changes in temperature. Realize that the birds will shift their comfort level from batch to batch by a good 10-15° F, depending on what they have been used to. Of course, the positive tradeoff is that if they have been used to heat, you probably won't have as much trouble with the birds as they get older if the heat wave continues. In regions where dramatic fluctuations between night and day temperatures are common, extra precautions would be warranted. If the nights are dropping in the 40s and it hits 80° F in the daylight, be sure to provide extra ventilation.

Generally, the birds do not like more than a 30°F fluctuation between night and day temperatures. Greater fluctuations cause stress. Chilling at night and burning up in the day are hard on a chicken. Since you can't do much to warm them up at night, the way to attack the problem is to make sure you can cool them down during the day. If your area is prone to this, you may want to hinge the vertical back of the pen so it can just flop up on top and let free flow air go through.

Being in the extreme western part of Virginia, right up against the mountains, we have dealt with a great variety of weather over the years. Be assured that with care, these birds are resilient. I don't want to unduly scare you with this weather chapter, but at the same time, this is one of the biggest variables in the production model and I don't want to mislead you.

If you've just moved the chicks outside, and a thunderstorm comes up, don't stay in the house waiting for the rain to stop. Don your raingear and check on them. We've lost plenty of chickens because we delayed checking on them until the rain was over. When those birds get wet, the time to deal with them is NOW, not 5 minutes from now.

It is common for us to have at least one or two week-long periods of rain, amounting to 5 inches, during the grazing season. Once surface runoff begins flowing through the pens, I make sure the pen has that pad of hay inside, then just let the pen sit until the rain subsides. The birds are much more comfortable staying on the dry hay than being forced out into the rain or onto the wet ground during the move, and it saves having to put in extra hay. One technique that we've used to get the birds on hay quickly is to put down hay and then move the pen onto it, rather than crawling around in the pen and putting down hay.

In northern areas, producers may not be able to have the birds outside quite as long as we do. We can usually put birds out by the third week in April. The first batch we may keep in the brooder for 4 weeks to make sure they are a little bigger before we put them out. We usually put the birds out at about three weeks. We always push the season on the first batch, but we want to be able to butcher them before we start making hay at the end of May. then we try to make our hay between the first and second batches. Our frost date is May 15. We can generally run the birds outside until into October.

In hotter areas, like Alabama, Louisiana, Mississippi and south Georgia, it may be wise to shut down the operation in the middle of the summer because of the heat. Of course, if you are in an area where year-round production would be possible, you could net $50,000 - $60,000 per year instead of half that in the six-month period like we would here. But remember that gives you no break and may severely impact your lifestyle. But if two months off are enough rest for you, perhaps you could raise for 10 months instead of 6, with a concomitant increase in income.

The important thing is to go with the flow. Different regions of the country will have different parameters. Don't fight Nature. You can't win. Stay within your area's limitations and you will be far more successful.

Weather is by far the biggest variable in the pastured poultry equation. It's like the girl with the curl in the middle of her forehead. When it's good, it's very good, and when it's bad, it's horrid. It accounts for the lion's share of everything that goes wrong when poultry is not confined to an environmentally controlled house. But the benefits far outweigh the risks by producing chicken that is fit to eat.

Chapter 22

Stress

The single most critical health factor is stress. In a way, animals are much poorer at handling stress than are people because they can't think through the problem. One bad experience mishandling calves, for example, will sometimes result in a weight gain reduction for as long as two or three months.

Minimizing stress must occupy the forefront of our thinking. Chicks must be treated like newborn babies. They must be warm and dry, and must have plenty of room in which to run - 25 square feet per 100 birds is plenty. If the area is too big, they will wander off in a corner and freeze, or lose the feed and water.

As they begin to grow, they begin to fill up the area so that by the time they are three weeks old, that 25 square feet is scarcely enough to give them room to run and play. One of the biggest temptations is to crowd them. Housing costs time and money. We know they are getting crowded, but we aren't ready yet to move them out and somehow we rationalize that an extra day or two in crowded conditions won't hurt anything.

Nothing could be further from the truth. It

is devastating. Just a couple of days of being crowded can set the birds back enough to be measured at slaughter. The number of birds in the group is very important. In the brooder, birds should be kept in groups of 300 or fewer. You can put 1000 in a group at the same housing density as the 300, and the smaller group will always outperform the larger one. Mortality is one of the biggest differences.

It is important to convey the stress factor to children, who love to go in and chase the chicks. Of course, all they want to do is catch one and hold it. But the stress that causes on the chicks is enormous. If a chick can be caught without disrupting the whole group, then certainly that is acceptable. My children learned as toddlers that they could not walk around in the brooder house; they could sit on a bucket I provided for them. They could sit and watch, but could not stir around. If a chick wandered close enough to grab, that was the only one they could hold. If they spoke, it was to be nothing above a whisper, lest they disturb the "babies." We must respect these little miraculous fluffy balls of life, and let them go about their natural business unmolested. And yes, we do believe in discipline, making children mind.

The same principles apply out in the pens. If you have a dog that comes with you to move the pens, he must sit quietly, away from them, while you tend the pens. I have no use for a dog that comes gamboling around the pens, whether I'm there or not. Such a dog should either be leashed or given away. You can't have an undisciplined dog or cat, and love your chickens at the same time.

Dogs and cats are predators of birds in nature, and if we domesticate them, we must control their behavior. The same is true with children. I get extremely sharp with visitors whose children run around the pens.

If parents become protective of their children, then I'll probably ask them to leave the farm. I don't play around when I know someone or something is panicking the chickens. We simply cannot tolerate it. The stress will have an impact, make no doubt about it. The stress factor takes its toll, no matter how innocent the play of the pets or children may appear.

Many people think moving the pens with a tractor is easier. Machinery and noise panic the chickens. The less noise, the less sudden movement, the less nonhuman intrusion, the better. When I walk quietly around the pens and tend the birds, they scarcely pay attention to me. That is the way to reduce stress.

Stress seems to be like pressure in that the explosive power increases geometrically, not arithmetically. Just a tiny change at the pressure "point of movement" causes dramatic differences.

For example, the number of birds in a pen has a direct bearing on performance. Of course, we want to put as many birds as possible in a pen in order to make the moving, feeding and watering more efficient. The same action is required whether the pen contains 20 birds or 100. The pens are as big as I can handle by hand. If you're going to move a pen, you may as well move as big a one as you can efficiently manage.

We have observed performance closely, with varying numbers of birds in the 120 square foot pens, to see how many birds gives optimum performance. We have found that under 50 birds, weight gain drops because they run around too much. But the problem generally is on the top end. The second year we raised birds, we put 150 in a pen. They grew up to about 6 weeks, and then just stopped. They continued to eat, but wouldn't grow any more. They were too crowded: stress.

By reducing the birds to 90 or fewer, we get the best performance. If we go to 95, slaughter weights drop. By increasing to 100, weight nosedives.

It seems that the 90-100 bird range is the critical area for crowding stress. Between 65 and 85, we have not been able to detect any difference. But every individual over 90 takes a toll. The point is that as we approach the top of the stress curve, every unit of increased pressure has a multiplier effect on stress. The beauty is that we need not back off much to drop the pressure off. Just a few birds at that threshold have a significant impact on performance.

Stress is obviously a critical factor in conventional confinement poultry production, where 10,000 or more birds are crammed into one house. It would be equivalent to a person living his whole life in a crowded football stadium. It's fine for an afternoon, but if you had to bathe there, eat there, change clothes there and sleep there, you would go bonkers. We were just not created to take that kind of pressure, and neither was a chicken.

I can't help but wonder how many diseases and epidemics could be avoided in the poultry industry simply by breaking up those birds into smaller groups and giving them a little more space. We have seen such dramatic differences by varying the numbers slightly that I can't help but believe the positive effects would be great for the factory farmers as well.

We must attack stress at all levels if we are to raise healthy birds. We cannot dilly-dally on this and expect them to cope with the pressure and go along their merry way. If they are uncomfortable, you will see it reflected on the scales. I guarantee it.

Chapter 23

Troubleshooting
Poor Performance

We raise Cornish Cross broilers to 8 weeks, generally slaughtering on days 56, 57, 59 and 60, and get a carcass weight that averages 4 pounds. Feed conversion runs about 2 pounds of feed to 1 pound of liveweight, or 1 to 3 at carcass weight. Carcass weight averages about 75 percent of liveweight. Mortality averages about 5-10 percent.

Remember that most industry feed conversion tables are for liveweight gains, not carcass weight. To ensure that you are comparing apples and apples, be sure to compute your conversion based on liveweight if you are comparing your performance to some Extension publication or industry chart.

Once you have raised as many birds as we have, you will experience the frustration of a batch from pullets or old hens. A breeder hen flock just coming into production lays small eggs. These pullet eggs produce small chicks, which the industry says need to be handled like "preemies." Regardless of what the experts may tell you, they will simply never perform like normal-sized chicks. Your mortality will be higher, the weight gain will be less. It's just one of those things that happens from time to time.

The opposite end of the breeder hen cycle is equally frustrating. When the breeder flock is just going out of production, and the birds are at the end of their laying time, the eggs are huge and produce equally huge chicks. This is the most frustrating, because when the chicks arrive you think you have a blockbuster batch. But what happens is a great deal of inconsistency among the birds. There will be some huge ones, and a whole lot of runts. In fact, some birds will look like quail, even at 8 weeks. They may look healthy, but they are tiny. The overall weight of the batch will be below normal. The main difference between a batch from pullet eggs and a batch from old hens is that the pullet eggs will produce consistently small birds and mortality will be higher than normal.

The batch from the old hen eggs will have generally a lower than normal mortality, but 1 or 2 percent of them will be these little quail runts. And overall, the birds will not be consistent. Some will be huge, but most will weigh below average.

The reason each batch isn't a carbon copy of all the others is that pasturing poultry inherently interjects the weather variable into the production model. If the weather is cold so that we must keep the chicks in the brooder facility for a few extra days, our weight will suffer. If we have a seeming monsoon while they are outside, performance will decline. Similar stress-induced problems occur if in the final week we have several 90-percent humidity, 100-degree days.

But beyond that, people who have tried to duplicate our system sometimes report mixed results, and I want to address the reasons for this.

The first and major problem is light weights. Instead of dressing out at least 4 pounds, they average 3 or 3.5. Except for climatic differences, there are several reasons this may occur. The primary one is ration difference, either in the grain part or the forage. If the birds are not moved to fresh pasture frequently enough, they will not stay healthy enough for either peak metabolic activity or nutrient assimilation. Chlorophyll is a detoxifier, and if they are not receiving enough fresh green material, the valves on this race car chicken may become gummy, so to speak. The forage supplementation as a percentage of total ration needs to be high enough to maintain extreme health.

Moving the birds every few days or so simply does not provide that level of green material and performance will necessarily suffer. A thin pasture sward will also reduce forage intake. Recall that there is a marked performance decline when the birds stay in the brooder beyond the fourth week. Greenchopping forage will not make up the difference. As soon as the forage is cut, vital antibodies and vitamins begin a significant decline.

The health that a totally new daily paddock affords the birds is of primary importance. If they stay in one place even 3 or 4 days, fecal pathogens will have time to incubate and cause damage.

The feed ration may be adjusted too much. For example, a typical problem is a feed mill trying to sell a "poultry pack" instead of the individual components I have delineated in our ration. The packs invariably contain salt, dicalcium phosphate, synthetic vitamins, electrolytes and other additives that militate against the raw, natural components like kelp, brewer's yeast and fish meal.

A prepared protein concentrate containing urea rather than straight meat and bone meal, for example, will not perform like the more natural product. Urea pulls vitamins A and D out of the body and deposits them, at toxic levels, in the liver. Not only does the bird lose the vitamins, but its liver gets clogged to boot. Feed mills typically balk at custom blending small amounts. We get nearly 4 tons at a time and have a feed mill sympathetic with what we are trying to do. It makes a big difference. I don't have the answer to the general problem of a "little guy" getting "big guy" service at the feed mill, but I do know that these changes they try to push onto the ration can have negative consequences, even when they say they won't.

For example, crippling is often viewed as a calcium problem. A typical feed consultant will offer a mineral blend to make sure those bones get enough calcium. But it is not the presence of calcium that's the problem, it's the assimilation, and that requires Vitamin A. That is why when we began adding alfalfa meal for Vitamin A and eliminated bone meal (calcium) from the ration, we saw our crippling nearly eliminated. At that time we were using a synthetic vitamin pack which had more units of vitamin A than the alfalfa meal, but the potency on these manufactured vitamins fluctuates, especially if they get hot. We eliminated the synthetic vitamins and suddenly got the calcium assimilation. Natural vitamins are much more stable. The point here is to beware when the conventional wisdom pooh-poohs what you are trying to do. Think for yourself, observe and don't quit seeking.

Another reason for performance to suffer is negligence. Litter dampness in the brooder facility is a prime example. That litter needs to be deep enough for good scratching and fluffing, and dry, dry, dry.

The quality of chicks from hatcheries can differ enough to account for some performance fluctuation.

The birds must be on full feed 24 hours a day. Inconsistent care can have a big effect. Letting the broilers run out of feed or water, moving the pens at 6:30 one morning and 9 the next and 8 the next, all upset the routine. Animals enjoy a routine, and the more it fluctuates from day to day, the greater the drop in performance.

Crowding can have an extremely detrimental effect, as can insufficient linear distance on feeders and waterers. Crowding in any of these cases shuts down performance fast. Those birds double in size several times in just 8 weeks. The little chicks that seem lost in the brooder facility the first day suddenly have no place to move within just a couple of weeks.

For optimal performance, the birds need aggressive care and observation. Certainly there are as many variables as there are people and places on the globe, but with proper care, pastured broilers should perform, without medications, hormones and synthetic feed additives, at least as well as their counterparts in factory confinement houses.

Chapter 24

Shortcuts

While I encourage innovation and a better way, I am also amazed at how many folks complicate this model when trying to simplify or shortcut our procedure.

We've been producing chickens in these pens for two decades. There are reasons why we do what we do. The program has evolved over a long period of trial and error. We want to protect you from some of the errors we've made.

One of the first shortcuts folks try is to eliminate the daily pen moves. This can be anything from not moving the pens as frequently to eliminating the pens altogether and free ranging the birds or confinement housing them. Two things invariably happen. First, the birds do not consume as much forage. Remember that consumption is directly related to freshness, and the pasture gets stale to the chicken far sooner than you or I can see it. The cream is quickly consumed. When forage consumption drops, so does feed efficiency and health. The bird's health is directly tied to the natural antibiotics and vitamins found in the green material.

Second, disease becomes more of a problem

because the birds are not daily moving off their excrement. The toxicity build up is real. In fact, some of the largest organic poultry producers in the country have gone bankrupt because of disease. You cannot have a spoke pasture, with loosely rotated paddocks, and expect the performance to be the same as unlimited, fresh forage. And you can't wait until the birds are two weeks from slaughter before putting them on pasture.

Remember that these birds will not range for food, so they will not get up and walk out of the building and come back in. They simply stay in the building and don't range. That is why we must take them to the pasture, providing it to them in such a way that they can consume large quantities without having to do much walking. We actually bring them and the plate together. Furthermore, these birds are too small, at three weeks, to handle a range setup. At that age, they have a hard time just walking through grass, let alone walking in and out of a building to find it. What normally happens, of course, is that the producer simply doesn't allow the birds access to pasture until maybe the last two weeks of life. But that won't do the job either. The disease problems are already there. The saturated fat is already there. Too much has been done conventionally to undo everything in the last two weeks.

For example, we have found that for every week beyond about 25 days that the birds are kept inside, we lose about a quarter pound of carcass weight at slaughter. I cannot emphasize too much the power of the grass, fresh air, exercise and sunshine.

Trying to shortcut the system with portable fences moved every few days, or a non-rotated pasture, simply thwarts all the positive things that the daily salad bar pasture provides.

And it's important to realize that moving the pens is not much work. I can move 2,400 birds, or 26 pens, in slightly more than 30 minutes. It takes roughly one minute per pen. Carrying feed and water takes an hour, because the older birds require servicing twice a day. All that servicing must be done whether the pens are moved or not. Moving the pens is not the time-consuming portion of the model. The time-consuming portion is in servicing the birds with feed and water, and in processing.

In fact, I must confess that I truly enjoy moving the pens. It is the most satisfying part of production, because I know how happy the birds are to be moved. When I see them put their beaks down and begin grazing that grass, chasing bugs, and then finally lounging down on the clean, fresh grass, they are the picture of contentment. I feel good, knowing that I have made them feel good.

Another shortcut folks invariably take is to make the houses bigger or heavier and/or move them with a tractor.

A bigger house spans more ground, and therefore does not fit the ground contour as well. Consequently, bigger pens make it easier for predators to get in. The larger the pen, the more ground it covers and the more irregularities it has around the bottom. And the pens need not be heavy. You do not need any 2 x 4 lumber. Think light and think bracing.

With the dolly, I can advance or stop easily, with absolute control, especially since I'm standing right there watching the birds. If one gets its leg caught under the trailing edge, I can stop immediately when I hear him screech, back up, and let him scamper ahead to join the rest of the birds. A tractor frightens the birds and gives you much less control when moving the pen. With a tractor engine running, not only will you not hear the chicken screech, but you will not be able to stop and back up as quickly or efficiently.

One reason many folks want to bring in a machine is that they build the pens too heavy. Please, please, do not build the pen heavy. Remember, it is not made of 2 x 4's and steel. It is light, dressed down 1 x 2's, 2 x 2's and aluminum, not sheet metal. The entire pen shouldn't weigh more than 200 pounds. If it does, it's too heavy to handle by hand, and that means it's too heavy.

Remember, these are chickens, not cows. It's hard sometimes for those of us who have built corrals and gates for cows or sheep to shift gears and build something this light. But lightness is everything in this case.

Some folks want to mount permanent wheels on the pens, thinking that the dolly is too cumbersome. Again, it simply doesn't take that much time. The entire process of putting the dolly under the pen takes about 5 seconds. It's as simple as a somersault to a gymnast.

Using steel instead of aluminum for the roofing adds a tremendous amount of weight to the pens. The extra weight can break the light lumber, which then requires you to use heavier lumber. The point of all this is to encourage you to stay as close to the model as possible. We are not trying to stifle innovation. We are trying to forestall the phone calls and letters we receive from disenchanted people who say, "we tried it and it won't work." It's amazing the things people can think of to change, and how big a difference these changes can make.

Moving the pens daily is crucial to success; pen construction is crucial to efficient daily movement.

Folks who come and visit the farm consistently marvel at how easy it all works. They can't believe that in half an hour we've moved more than 2000 birds onto fresh pasture. But it works, and it's not hard work. It can be done fast or slowly. It doesn't have to be done in an instant. It's a perfect low-stress job.

Now to the other part that's too much work: the processing. Many folks love to produce the birds, but they don't want to touch the processing. Sure, it's work. That's the hardest part of the whole program. But that is where the money is, and it is also the only way to ensure that your good chickens stay good all the way to the consumer. As soon as those birds go through the transportation stress to a custom slaughterhouse, and get exposed to pathogens from confinement-raised birds, the toxic sanitizers and electric killing procedures, your good chickens have just been adulterated.

In order to ensure the integrity of your product, and to get the full benefit of the value-added concept, I encourage you to follow through with the processing. Remember that the processing only takes a few mornings a summer. With this model, at the 10,000 bird level, you will only spend two mornings a week processing. Six months of the year you won't spend any time processing. That's when you rest, read, and do other things. Anyone can sprint for awhile. Sure, we aren't built for a marathon every day, but then neither are you being asked to run a marathon every day.

There is no free lunch. The simple fact is that you cannot enjoy the lucrative nature of this enterprise without following through to the nasty part of butchering. The bottom line is built around many details, all of which have to fall into place and all of which are important. As soon as we start tossing out items, the bottom line begins evaporating, too.

I encourage folks to come for a visit. See for yourself how it works. We have no secrets and no hidden agenda. All we want is for you to be successful and for the customer to be happy and healthy. Let's not get bogged down trying to reinvent the wheel, but learn from each other. Refinements are welcomed.

Chapter 25

Seasonality

Our entire poultry model assumes a seasonal production. This flies right in the face of conventional thinking. After all, people eat chicken year round, not just in the summer. Ah, the wonderful technology of freezers.

The more we study the advantages of seasonal production, the more satisfied we are that we have made the right decision not to be suckered into the demands of year-round production. Nature provides seasons for a reason. Each has its assets and liabilities. Producing in sync with the season provides the greatest profit. For example, seasonal grass-based dairymen have learned that the non-producing time should be winter in the north and summer in the south. Either extreme cold or extreme heat reduces good forage growth and cow comfort. Off-season production requires stored feed and housing the cows either for warmth or shade.

While here in Virginia we would stop producing chickens in the fall, folks in the deep South might stop their season in the late spring, electing to sit out the hottest, driest part of the year.

Weather is the only real factor determining what season to use. That affects both forage growth and minimal comfort. By going with the season, we can produce the best quality product at the least cost.

As soon as we produce out of season, costs escalate and performance drops. The birds are not as comfortable, forage is not as good, disease is more prevalent, and we end up spending more on heat or shade, as well as mortality losses. The bottom line doesn't look nearly as good as it does in the better season.

Our Virginia seasons in the Shenandoah Valley typify much of the United States. With a median winter temperature of 35° and median summer temperature of 75°, we are certainly in the mainstream. We usually have a week of 0° in the winter and a week of 95° with 90% humidity in the summer.

Reasons abound for seasonal production. First, we can't raise the birds outside in the winter. Three-week-old chicks just don't do well when the thermometer drops below freezing and the snow begins to fly. Clearly we would have to raise the birds indoors in the winter, providing them additional heat at a healthy additional cost. Furthermore, fresh air and sunshine would be a major problem, and fresh green forage would be unobtainable. The birds would be sicker, would not gain as well, and processing in the cold is no fun. Indeed, a winter-produced bird would be completely different than a summer-produced bird.

Even if we assumed the extra production cost, the saturated fat would be far higher and the meat would not taste as good. We would lose the detoxifying benefits of the fresh grass and

... well ... you get the picture. I think we can agree that the product would simply not be comparable.

To make the same income per unit for off-season production, assuming the poorer quality bird didn't matter, we would have to charge at least 50 cents more per pound, which would put us way beyond store prices and out of the average consumer's reach. Because we would be producing on a small scale similar to the mass-produced birds, we would lose our competitive edge. That is not where we want to be.

In the big picture, spring and summer is nature's bird-raising time. That's why chickens lay eggs in the spring. Nature must intend birds to be born and raised in the spring and summer, because that is when the increasing daylight stimulates egg production. After all, egg production is primarily for the benefit of producing young, and only secondarily for the all-American fried egg.

We have found that hatchery eggs are better in the spring than in the late summer. The chicks arrive healthier, and they just perform better. By far our best batches are the middle batches. The birds hatched in March are from eggs laid in February, which is about a month or two ahead of the spring lush. The best chicks come from eggs laid in late spring and early summer. Incidentally, that is when the quail, grouse, meadowlarks and robins are reproducing. Nature knows best.

That is also when eggs are the most plentiful. Most hatcheries close down in the fall and winter, simply because it is too difficult to ship chicks in the cold, and because it is too difficult to get a supply of hatching eggs.

Clearly we are in sync with nature by limiting our production to the times when chickens naturally want to lay the most and best eggs.

Another asset concerns the manure generated in the field. Any time manure is spread on dormant ground, most of it is lost. The time to feed the soil is when it can assimilate the material, or capture the highest percentage of nutrients. That time is when it is vigorously growing green material, when biological activity is at its peak. And that time is in the summer.

The very time when we need the grass most for the grazing cattle, and when the grass can best utilize soil nutrients, is the time we are stimulating growth by putting on the chicken manure. If we put it on at any other time we could not utilize its full potential.

Next on the asset list is the manpower component. This includes both the producer and the consumer. The chickens increase the summer workload, to be sure. But this is when the children are out of school and the days are longer. From a labor standpoint, this is the perfect time to add a component to the farm's chore load. If you need help processing, teens can be brought in to help because they are out of school ... Summer is the ideal time to add work to the farm.

In like manner, consumers are more relaxed. The hectic school schedule has given way to summer recreation. There is time to bag and freeze chickens that is simply not available when it gets dark at 6 p.m. and PTA meetings, football practice, cheerleading tryouts and a host of other extracurricular activities occupy a family's time.

Who wants to drive out to the farm during a freezing rain, slog around in the mud, see the leafless trees and stick their hands in icy water? But in the summer, when the fields are pastoral green and the garden is in full production, when blue skies and gentle breezes beckon, a trip to the country is certainly in order. In fact, for the consumer, it's an enjoyable outing and becomes part of the routine for laying up winter's provisions.

Which brings us to the problem of winter. What about eating chicken in the winter? We simply rely on the modern convenience of the freezer. To be sure, many customers have had to buy freezers because of us. We've joked in the family that we ought to start selling freezers. Long before freezers, the idea of laying up a store for winter was normal practice. Capitalizing on nature's bounty during the spring, summer and fall to carry us through the winter via some sort of food preservation has been done throughout recorded history. Seasonal production and storage is not new.

In nature, poultry is a lighter, low-calorie meat, while herbivores produce a heavier, fatter meat. We should eat poultry in the summer (along with fruits and vegetables) when we are trying to stay cool, and eat beef and the starchy vegetables like potatoes and winter squash in the winter when we are trying to stay warm. Grass fattened beef is best when the animals are slaughtered in November and December. Amazing, isn't it? Nature truly knows best.

I encourage consumers to take the food storage discipline to new heights by not buying "fresh" tomatoes in January, but rather to patronize the local tomato grower in July when he's overrun with the delicious vegetable, and can them for winter use. That makes more sense from every angle: socially (the dollars turn over in the local neighborhood instead of being exported); economically (the tomatoes are cheaper in season); and environmentally (little energy is necessary to transport the tomatoes from producer to consumer). They also taste better. I realize the notion of the twentieth century family devoting time and energy to canning and freezing could be perceived as an archaic idea, but truth often comes askew of conventional ideas.

It is amazing to me how many people can spend big money and big hours figuring out how to have a successful vacation to Disney World, but can't devote any time and energy to ensuring a healthy, socially progressive diet in the winter. The fact is that often we major on the minors and minor on the majors. I believe this objective for the modern household is consistent with Wendell Berry's writings on home economics, and with Wes Jackson's writing on modern information implosion.

When we look at the big picture, the cycles of nature are here for a reason. As we begin tapping into them, and using them for our benefit, positives occur in many areas: social, economic and physical. To fully utilize the season requires forethought, discipline, and seasonal hard work. Solomon pointed out that there is a season to plant and harvest. As soon as we refuse to plant or refuse to harvest in season we pay a price.

If everyone adopted this seasonal posture, what would happen to the poultry industry, the processors, in the winter? First of all, the workers could do other things; secondly, the breeder hen eggs could be sold through conventional channels to supplement the drop in layer production (which then could be seasonal and non-light-stimulated), and the processing plants could open their doors to some fresh air and sunshine for a sanitizing kiss from nature.

Our requirement here, however, is not to answer the question, "What if everybody did it?" I do not believe it will happen. Our purpose here is to show the benefits for you, the grower, and you, the consumer. Certainly, we would be remiss by not looking at the bigger picture, but in reality we know it won't happen and we can function as a tiny subset of the poultry industry for a long time. These are excellent marketing points for promoting your product.

From a marketing standpoint, the seasonal approach offers some benefits because our newsletter comes out in late February, when most people have practically exhausted their supply of chickens. Invariably, consumption increases when the cook has a stockpile of birds in the freezer. They appear cheap because they are already paid for. In other words, psychologically no cost is attached to them because they are already in the freezer. As a result, people eat more chicken, run out in late winter, try one from the store, gag on it, and then receive our newsletter. "Ah, what a relief," they say, adding, "I don't ever want to run out of those chickens again!"

It is like a breath of fresh air, and they assure themselves that this time they are going to order plenty. It certainly takes some planning and foresight on the part of the customers who buy their chicken seasonally, but usually within one season they learn to think ahead. For sure, some folks drop out. They just can't handle forethought. For them, we say good riddance. You know the old saying about being unable to please all of the people all of the time. It is true. Society is full of losers and winners, and that will never change.

Remember, we are not trying to sell to the masses. We are trying to sell to folks who appreciate good food at a reasonable price and who get with a program that enables them to have good food at a reasonable price. Otherwise they can have good food at an unreasonable price or bad food at a reasonable price. I don't see much middle ground.

Of course, the biggest benefit of seasonal production is the rest period for the producer. Any job, no matter how small, can become drudgery when it must be repeated forever. And a big job looks smaller when you know there is a finish point. When there is an end, a change of pace, we can anticipate the coming of the first chicks and the going of the last ones, and enjoy the excitement on both ends. About the time we are emotionally running out of steam, we're getting the last batch. Seasonal production is where the bonuses are: financially, emotionally, qualitatively and environmentally.

Chapter 26

Solving Your Own Problems

A pioneer is on his own. Forging beyond
traditional boundaries, whether they be academic, philosophical or practical, carries with
it inherent risks. The greatest risk is that we
must solve our problems ourselves, with neither
the encouragement nor the knowledge of others who
have gone that way before. So it was in our fifth
year of raising broilers unconventionally that we
had to overcome a terrible problem. What follows
is a recap of our observations, our thinking and
our experiences.

Our goal in recapping this event is to
encourage other pioneers to conquer their unique
problems. And to illustrate that conventional
thinking does not apply to pioneers: it just
can't adapt.

In August we received a batch of a thousand
day old Cornish Cross chicks. The summer had been
successful thus far, with little mortality and
good carcass weights. These chicks, smaller than
normal, were hatched from pullet eggs. The
breeder flock was just coming into production,
and their small eggs made small chicks. We were
assured that the genetics were fine, but they

needed some extra care.

They started off fine and on about the ninth day I moved them out of the brooder house into our portable, floorless cages on pasture. Two days later, several birds showed lameness. The hyped up breeding in the meat chicken industry has brought with it leg problems, a major plague in the industry. Subtherapeutic antibiotic administration reduces joint swelling and reduces the crippling.

Since we do not use antibiotics, we fight lameness by adding alfalfa meal to the ration for organic vitamin A (aids in metabolizing of calcium), being fanatically meticulous about moisture in the brooder house (keeping the floor dry), splitting the soybean meal component of the ration in half so the other half is roasted soybeans (keeps the vitamin rich oils), and adding probiotic (Lactobacillus acidophilus) to the ration.

Just when we put this batch of birds out in the field, the weather turned cool and damp. Immediately I attributed the lameness to that. Over the years we had come to believe that in small birds lameness was incurable. We disposed of those birds quickly to reduce their suffering.

The next day, which was day 13, more chicks became lame. It peaked at about day 20 and tapered off until day 28, when we found no more cripples. At that point, we had lost about 30 percent.

Another batch of chicks came, similar to the earlier one, and they started the same thing at day 12. I kept these in the brooder house, however, to make sure they stayed warm and dry. When the same thing began happening to them, I knew it had nothing to do with the weather. At day 14, I took cripples from both groups to the state laboratory for diagnosis. We couldn't afford another loss of 30 percent.

I sat in the brooder house and watched the lame birds. Their problems seemed to begin with foot paralysis. Their toes would curl under; this would throw them off balance, and then they would rock back on their hocks to move around. As it advanced, the paralysis would affect the entire leg until finally they would be dragging both legs behind them as they flapped their wings to move around. It was a most pitiful sight and one I shall always remember. Nothing was wrong with their appetite, either for dry food or water. Their eyes remained bright and color was good.

Pulling my *Merck Veterinary Manual* off the bookshelf, I began leafing through the poultry section, looking for anything that seemed to fit. By the time the state lab spent a month on its report, I knew its information would not be in time to spare my losses.

I came upon a heading titled "curled toe disease" caused by riboflavin (Vitamin B12) deficiency. It described this malady as "characterized by a tendency of the chicks to rest on their hocks and a slight curling of the toes." It fit perfectly. Immediately I grabbed Adelle Davis' *Let's Eat Right to Keep Fit* and looked for foods high in riboflavin. It listed greens, but especially organ meats, including beef liver.

There was no time to lose. I sent Daniel, our son, to a patch of ragweed with instructions to start pulling off the immature green seeds. I went to the freezer and pulled out a pound of beef liver, cut it up as small as possible, and took it to the chicks. They swam in it! They rolled in it! They absolutely went crazy. Then they devoured the ragweed seeds.

I felt optimistic. Perhaps that was why the first batch stopped going down a week after going outside, I reasoned. The weather had not caused it, as I had thought. Apparently the green grass stopped the lameness. The grass was not nearly as potent a riboflavin source as liver, but nonetheless put a stop to it in a few days. The first batch was 5 weeks old by now, and doing fine.

We continued feeding the ragweed seeds and liver to the second batch. They kept devouring it and it stopped further lameness.

I quarantined the ones who were already lame and in a few days they were healed and went back with the main bunch. Only two never came out of it. We discontinued the liver after a few days and there were no recurrences of the lameness. In the first batch, we lost 30 percent. By feeding ragweed seeds and liver at the first symptoms, we salvaged the entire second batch.

We have not had this problem since putting brewers' yeast in the ration. We feed liver only to the few who develop leg problems. They stay in the hospital pen until we can move them back outside.

The reason those two batches of chicks developed the problem is subject to debate. My theory is that because it was extremely hot during the time the parent pullets were laying the eggs, and the synthetic vitamin paks in the bulk feed tanks were breaking down. This has happened in commercial poultry operations before. The reason mine showed up with a problem was that I didn't feed antibiotics to block the infections in the nerve tissues of the legs.

Another possibility is that these extra

fragile chicks required a higher B vitamin dose than normal-sized chicks. The deficiency showed up more acutely than it ever had. By adding brewers' yeast to the ration, we hope to never have this problem again.

What of the state lab report? It arrived about six weeks after I took the sick chicks in for diagnosis, far too late to have solved the problem. The official diagnosis was Marek's disease, which I considered ludicrous. This disease normally attacks older birds 12 to 24 weeks of age; ours were chicks. Furthermore, it is highly infectious and communicable, which is inconsistent with the fact that it stopped in the first batch. Finally, in Marek's, paralysis affects the entire bird, not just the legs. Staph infection and *E. coli* were isolated from leg nerve sheath cultures.

The lab advice was to throw antibiotics at them, but the prognosis was not good even with that. The advice came on a standardized listing of antibiotics with accompanying chart marked whether the infectious strains were susceptible or resistant to the drug. This is standard fare for poultry producers.

More amazing than the fact that the lab misdiagnosed the malady was the fact that the birds healed themselves of nerve tissue infections without antibiotics when their diet changed to the proper level of nutrients. Today's conventional wisdom finds this hard to accept. After all, infection must be fought with antibiotics, right? We all know that to treat infection with nutrition is quackery. This certainly offers an important lesson for the human deaths resulting from eating *E. coli* contaminated meat. The problem is directly linked to nutrition and production models.

We've raised many broilers since that time, with some batches showing less than 3 percent mortality. The lesson here is that conventional thinking imprisons creative thinking by limiting it to traditional thoughts. As Allan Nation says, when the only tool you have is a hammer, every problem looks like a nail. The state veterinarian, in this case, could not think as creatively as I could, because the only tool he had was an antibiotic. Boldly think for yourself, press on in the pioneer mode, and take establishment wisdom with a grain of salt.

I get reports from folks all over the country who go to the poultry science professors at their state land grant colleges for advice about raising chickens this way. The normal response is "that's impossible." In fact, one prominent poultry science department head walked out of a presentation I was giving, declaring that "humanity would die of disease if all these chickens were put out in the air where their diseases could infect people." I wonder how in the world he thinks people and birds existed before confinement factory housing came along.

The fact is that the paradigm to fighting disease is to throw antibiotics at it. The poultry production paradigm is to put thousands of birds in a horribly filthy place for a few weeks and then take the survivors off to a slaughter plant.

Because the pastured poultry model does not fit within conventional paradigms, folks operating within those parameters cannot understand what we are doing. You and I, as poultry graziers, do not fit within their framework. It's amazing what nonsense can come from the mouth of a Ph.D.

Just remember that you can solve your own
problems, and be willing to entertain any notion
as being the cause or solution.

MARKETING

Chapter 27

Marketing

Our marketing system is as unconventional as our production system. And really, who would want to raise such a good chicken only to have it enter conventional channels to conventional people who would have only conventional appreciation for their food?

Sometime in February we send out a newsletter containing an order blank and a self-addressed envelope. The letter levels with customers about our concerns, needs and dreams. [See Appendix B for examples of newsletters.] The order blank articulates price and when the different batches will be available. The customer marks how many birds he wants per batch and sends it back in the enclosed self-addressed envelope. Everything is first come, first served.

Sometimes it takes people a couple of years to begin planning for their poultry needs a year in advance. We purge the customer list every year. Anyone who does not order for a year is purged from the file.

We raise the batches about 3 weeks apart during the summer, beginning the first chicks the end of March and processing the last bird around the first of October.

We order chicks from the hatchery to fill the orders. That way we do not take the risk of selling the birds after we've raised them. They are sold, or contracted for, in writing, even before we receive the chicks.

About a week or 10 days prior to processing, we call all the customers who have ordered birds for that batch and give them the option of two, three or four processing days, depending on how many days it will take us to process the entire batch. Customers may come between 1 and 5 p.m. on the day of processing.

In all the years we've been involved with pastured poultry, we've not had a problem being "stood up" by no-show customers. The chickens are superior enough that customers come for them like a religious observance, like a pilgrimage. I cannot overstate the loyalty of these customers, most of whom bring their children, visit the brooder house and pick up the baby chicks, or hike to the pasture and watch the bigger birds grazing in the field.

The bond between producer and consumer is incredibly strong with this direct marketing approach, and word of mouth spreads the news like wildfire. The biggest question we receive and the area of greatest concern is: "How do you get customers?"

We farmers have become so far removed from the consumer that the thought of actually selling a product is completely foreign to most of us. Hidden behind this question is the unspoken one: "It worked for you, but it won't work for me. People are different here."

213

I disagree. While this question and this concern seem to plague nearly everyone contemplating this venture, it is clearly the easiest hurdle, based on feedback we receive from those who have actually begun producing and processing birds. The folks who have begun are consistently incredulous that the birds have been so easy to sell. It is the biggest relief and the biggest surprise.

Realize that with our bent to export agricultural commodities from the U.S., we have lost sight of regional food production/consumption, and our neighbors as customers. How many people around you are growing chickens? That neighbor down the road would love to eat good chicken for a change. Give him one.

During a speaking tour in Nebraska recently, I was especially grilled on this issue, since the population is more sparse there than it is here. But at two meetings, I found people who were producing up to nearly 2,000 broilers and reported that demand was excellent. I calculated that it would take 100 of us just to supply the city of Lincoln.

Certainly I appreciate that everyone can't do it. But instead of jumping to number yourself among those who can't, why not be the first to do it and enjoy the benefits of being the trail-blazer? The point is that many people could do this who aren't. Certainly the opportunity is not for everybody, but if we consistently delayed doing something until we could prove that everyone could do it, we would never do anything. The race is never to the pack; it is to the fastest, the first, the frontrunner. Just because the whole pack can't receive a trophy is no reason to cancel the race.

Realize that acquiring customers begins one person at a time, one chicken at a time. Then it dominoes from there. And remember that one of the biggest dangers in the program is to grow too fast. Raise a few birds for yourself and give some away if they are good. Then raise a couple of hundred. We did not hit 1000 until our fourth year of production. Of course, we didn't have the benefit of a manual, but the point is still strong: start small and build one person at a time.

You must be your own biggest fan. As an entrepreneur, with government regulations to hurdle, it is not enough to be mildly interested in your product. You must be just plain passionate about it.

Get familiar with the mechanics of the production and the processing. If the chickens do well, give a few to your friends and let them know you will be glad to raise them some the following year, or later in the summer. The technique of giving samplers is used by businesses every day because it works. Starting this small will allow you to go through the inevitable learning curve required for any new enterprise. And losses will not set you back too much financially.

Once you are confident of your production and processing capabilities you can become more aggressive in building your business. Techniques are as varied as the people who begin a business.

We did a slide program and offered free birds to people who recommended new customers to us. One friend hosted a gigantic chicken barbecue for all her family, friends, neighbors and anyone she thought may be interested in chicken worth eating. By the end of the picnic, people were signing up left and right to get on the chicken list.

Clay Ham, in Texas, is in an extremely rural area. He contacted a couple of food coops and gave them some chickens. He had almost 250 families as customers immediately, because once the coop bigwheels tasted his chicken, they wanted Clay's chicken exclusively. He also hosted a homeschool shindig at the farm, serving his barbecued chicken. It was a smashing success and garnered him many new customers.

You may want to print up a brochure about your operation that you can distribute to interested parties, or give out at your state's organic farmers association meeting, or at meetings of ecology groups. A booth at the local county fair, with bite-sized samples of chicken, could spread the word. A float in some area firemen's July 4th parade could generate interest.

Be creative. Anything can work as long as it is true and done in good taste. All you need is for people to eat the chicken one time. That's all it takes. Whatever it takes to accomplish that in a moral and legal way is worth pursuing.

Remember that there is real danger in growing too fast. We've found that word of mouth has brought us about as many new customers each year as we could handle.

Be patient, do a good job, and people will respond accordingly. There is an old saying about a person who has a good product at a reasonable price never wanting for customers. Even in our throwaway society, people are still interested in good food at a reasonable price. Just jump in, raise a few, and get started.

This marketing approach allows us to focus our attention on what we do best--growing the world's best chicken. We have customers who ask us to deliver, or ship, or bag, or freeze, or cut up, or any other amenity you can imagine. Every time we try to oblige, we find our time and energy cut short. And meanwhile, our attention to production is diverted and we don't do as good a job in that area. It is best to "just say no" to these amenities. Some people would like you to cook it and fork it into their mouths.

Unconventional food must also be marketed unconventionally. Otherwise, the economies of scale force the unconventional food to be too highly priced for the mass market. As price rises the market pool shrinks. The best way to compete is by marketing unconventional commodities unconventionally.

We have eliminated all the amenities, realizing that if people who spend hundreds of dollars on a personal vacation fling can't make the effort to come to the farm and acquire the world's best chicken, then they don't deserve it. Some people will simply not respond to the idea of taking control over their own food supply. But many will. These wonderful folks will help you build a thriving pastured poultry enterprise.

An entrepreneur can easily sink in red ink trying to float everyone's little desire. The temptation is strong, because we want to please. We don't like being hardnosed. But some customers are simply a liability and not an asset. The relationships built with those who do "get with the program" more than compensate for the losses of the prudes.

Not every sale is a profitable sale, and we must weed through people until we find those who are good for our business.

Remember that as a small producer, we must resist the urge to do the things that big operators do well. Big producers/processors excel at materials handling, packaging, mass marketing and transportation. Small producers excel at relationships, high quality and integrity. Do not get sucked into competing where economies of scale ensure your failure, but rather do only what you can do better than the conventional system.

A local health food store sells organic chicken for $7.04 per pound. It's deboned and air freighted from California. Is it any wonder ours at $1.25 is easy to sell? That is a graphic illustration of where a business with noble goals got sucked into competing with Madison Avenue. That price makes a mockery of alternative food, and simply fuels the derision from the conventional sector. Enjoy the benefits smallness offers. Believe me, factory farming is a hard taskmaster.

By tapping into the consumer's car, refrigerator and time, you need not charge for those services, and from a wholistic viewpoint, that is an overall more efficient way to go anyway. You need not add another layer of duplicate capitalization in the food chain. It might not be the best for GNP, but it certainly is the best for the environment and the economy in the good sense of the word. And it's quite efficient when measured in calories of total energy required to get a pound of food on the table.

We unconventional food producers need to see beyond the commodity to changes that can truly bring the rural and urban communities into partnership on a regional level. We want consumers to become aware of where their food comes from and how it is produced. And that can't occur when Virginians buy California-raised poultry - or lettuce, for that matter. To facilitate the broad changes our culture must go through, we must concentrate on our localities. Then the big pieces will fall into place. That is why I'm not interested in building an empire or growing millions of chickens. I would much rather see a farmer in Richmond take our Richmond customers, and one in northern Virginia take our Washington, D.C. customers, so that local folks could develop a tie to the land and their food, and begin that awakening that happens as we get acquainted with brand new worlds.

Just because a California producer can sell chicken in Virginia doesn't mean he should. Just because a market exists for something doesn't mean that it's necessarily right to sell it. Markets are bound by what is moral, ethical and the best for God's total creation. Any other goal, any other desire, is not noble enough to warrant our time and energy.

Chapter 28

Relationship Marketing

In light of the insatiable human desire for empire-building, perhaps it would behoove us all to back off and examine the merits of relationship marketing.

While economics lessons generally laud practices that increase business size, business volume and annual quantitative growth, equal time should be given to the notion that numbers do not equal profits - or happiness. I was fascinated recently with an article in one of those freebie airline magazines about time and work. Twenty years ago, experts were predicting a 30-hour work week, more leisure and a Jetson's lifestyle of button-pushing and console monitoring. The fact is that people are working longer hours per week at stress levels that eclipse those of just a few years ago. We are in fact working more, not less.

While not every farmer can sell products at retail prices directly to the consumer, many more can than can't. By selling only to those who will come to the farm, the producer will inherently remain small compared to empire-builders who pack, ship and advertise all over the country. While we do not have thousands of customers, our several hundred are more than enough to provide the income and lifestyle we desire.

The first advantage of on-farm direct marketing is consumer education. In practically every agriculture publication I receive from the conventional crowd there is some article about those stupid consumers who don't understand this thing or that thing. Consumers are considered ignoramuses, constantly trying to "do the farmer in" by supporting ludicrous social and legislative policies. But instead of opening our farms up to them, we post big **NO TRESPASSING** signs at the entrances, especially on poultry farms. How can we educate consumers, how can we build a bridge to them, how can we garner their trust if we shut them out of our farms and call them names?

Consumers are just like producers. They are inquisitive people who reciprocate treatment in kind. They are sincere, honest folks who can't be held responsible for their ignorance, and who need to be pushed to learn something. Don't miss that last phrase. They need to be pushed to learn.

If we as farmers continue to market in a way that allows consumers to be uneducated about their food, are we not partially to blame for their ignorance? The fastest way to educate consumers is for farmers collectively to quit selling their wares through channels that inherently distance the consumer from their place of origin. I realize this idea has practical limitations, but it is still a valid concept and one that farmers can do their best to adopt, one at a time.

How in the world can the city consumer know what the farmer needs? He won't read agriculture publications. As the farm community continues to shrink and the consumer crowd grows, as the gap widens and mutual understanding decreases, how can the farming community possibly expect to hold its own? It will not. It cannot.

Is it too much to ask consumers to learn? I do not think it is too much to ask that a family that will routinely spend $1000 or more on a Disneyland vacation be encouraged to drive into the country once in awhile to see how their food is being produced. Not too long ago farmers peddled their wares at roadside stands, farmers' markets and as vendors door-to-door in the cities, engaging customers one-on-one in educational, bridge-building conversation. That has changed for the most part, especially in livestock.

And what we have now is agriculturally illiterate consumers with no link - mentally, emotionally or physically, to their food. We farmers, fiercely independent and individualistic, must certainly share the blame because we have turned our products over to Madison Avenue for marketing.

Probably the most distasteful part of our direct marketing is processing chickens. It's a nasty thing when we're running full speed. Blood drains, the scalder emits a steamy, wet-feather aroma, guts roll out on the eviscerating table. It's real exciting, something we should keep hidden from the consumer, right? WRONG! Many of our customers come early to see, and show their children, how we dress the chickens. Some customers come and help, just to be involved with it. Sometimes we feel like Tom Sawyer painting the fence ... getting all his buddies to pay him to do the work.

Consumers are starved for a link to "grandpa's farm," a link to their food supply. They can see that the animals have no understanding of the slaughter process like humans do. This keeps them from becoming animal rights activists, who say a fly is a dog is a human. Anyone who has watched animals being butchered knows animals are not human. Sometimes children - boys especially - like to take the knife and slit a throat. Talk about hands on education. Children learn that chicken doesn't come from a plastic bag.

This educational process is absolutely essential to keep the cities from running over the country. Mutual appreciation, instead of mutual mistrust and fear, is truly a goal worth pursuing. For sure the consumers will not initiate the process. Farmers, one at a time, must try to regain the ground lost.

The second advantage of remaining small has to do with product quality. Dad always reminded me that the biggest temptation of any growing business is to compromise product quality. Cutting a corner here and cutting a corner there add up much faster when more corners are involved. When only a few units are produced, a dime here or there doesn't add up to much. But many units, times a dime, can mount in a hurry.

If all the large alternative agriculture producers who have gone belly up in the last five years were tallied, it would be mind-boggling. The reasons for failure are myriad. Some were due to bad financial decisions. Many were due to disease/pest problems as the operation grew. This has been especially rampant in the animal sector of alternative agriculture. One lamb doesn't make a flock. Pets are always prettier and more productive than commercial, working animals. One chicken in the backyard is fine, but put that chicken with 10,000 others in a building and see how it fares.

223

Others have failed because their cash flow could not keep up with the initial investment. Paul Harvey calls this "overrunning your head-lights." A small business, perhaps initially supported by outside (off-farm) dollars, can grow slowly and capitalize internally on its profits. The bugs can be worked out little by little. But in a highly indebted operation, there isn't time or money for fine-tuning. Typically, the businessman becomes a fireman, putting out fires. He's never really in control; the business controls him.

That thought reminds me of the common response in our area when a contract poultry grower is asked how things are going: "Well, it's making payments." That's not good enough. It needs to make a salary.

If size or magnitude were the way to wealth, big farms wouldn't be sending out liquidation notices so frequently. The ability to maintain control of the livestock and control of the health, vigor and vitality of the product is much simpler when the enterprise is small.

The third advantage is customer loyalty. It is axiomatic that customer loyalty grows as the link between producer and consumer shortens. If all the consumer knows is a label, then it's fairly easy for a brighter, snappier label to win him over. If all he knows is the advertisement, a sharper ad with a little more pizzazz can win him over. But when he knows the producer personally, suddenly the relationship element comes into play and nothing short of total disenfranchisement can pull the customer away.

The relationship builds those emotional and physical ties that are not easily broken. This loyalty cannot be measured in dollars and cents.

224

It can't be put on a balance sheet. The cornerstone of a successful business is repeat patrons. A business cannot survive if customers don't continue to purchase. The customer turnover rate is directly related to the loyalty rate. The greater the loyalty, and hence the satisfaction, the higher the likelihood of repeat business.

I remember well a customer telling another person who was querying him about what we fed our chickens. "I don't know. They could feed them diesel fuel. Whatever it is, it must be good to make them taste the way they do." Now that's loyalty.

When the consumer is removed from the producer, and is not knowledgeable about the farming practices, a seed of doubt regarding product integrity, business ethics and the like can easily send the customer looking elsewhere. But if he has visited, seen first-hand, and participated in the farm operation, he will have the knowledge necessary to withstand some unfounded rumor.

I can't imagine why a business with an educated, loyal local clientele would want to trade that off for a nameless, faceless, capricious clientele from the four corners of the globe. Perhaps that is why many good small businesses grow up to be bad big businesses. And in the transition, many fail.

Distance, both in time and knowledge, builds an inherent gap between consumer and producer. Producing for a local or regional market, while not an option for everyone, is certainly an option for many, if not most. That bridge can and should be built. It's worth it.

The fourth advantage is lifestyle. I know the world is full of high rollers. Some people seem to thrive on being big. Dad used to call it "being born with a big auger."

But for most of us, the price of size gets paid in ulcers, loss of our children to things we don't have time to teach them not to do, marriage breakups and a host of other lifestyle traumas. As the empire builds, so does the pressure. Financial pressure, the work load, dealing with employees - all of these take their toll on our lifestyle.

We all know that money doesn't buy happiness. Some of the most unhappy people in the world are millionaires. Wealthy people often worry more about keeping their millions than the rest of us do about acquiring money.

I think one of the biggest differences between the pressures I encounter as a small potato and the pressures encountered by big potatoes is the amount of control we have over the situations that cause pressure. Big pressure for us is if a deer runs through the cross fence and the cows get into the next paddock. Or perhaps some cold, rainy weather requires that I go out and put some hay paddies in the chicken pens to keep them warm and dry. Big pressure for me is flicking the switch on the chicken picker and it doesn't go, or having an element burn out in the scalder. In either case, I need to replace the items, which causes downtime in a hectic morning, but that's the extent of the crisis.

As traumatic as each of these may be at the time, they are nothing compared to labor relations trouble among the employees, or inability to finance the Friday payroll check, or a truckers' strike that keeps you from being able to get your product to customers. These problems, because they result from interaction with people and other businesses, are much harder to control. They can't be "fixed" as easily as things under your direct supervision on your family farm.

This is the element that has led financially troubled farmers to suicide: when they realize that the financial picture is completely out of their control. As long as we are in the driver's seat, we can take all sorts of rough driving. But when we have no control over the vehicle and some yo-yo is behind the wheel and there's no way to escape - that's when lifestyle goes to the dogs.

No one can escape from the pressures of life, whether they be financial, emotional, physical or spiritual. But the chances of affecting those pressures, of dealing with them, of solving those problems, make the difference between an enjoyable lifestyle and a terrible lifestyle.

We all have 24 hours in a day. Within the parameters of our gifts and abilities, we can choose how we will spend those hours. I have no desire to complicate my life with an empire. I'd love to see thousands of independent farms serving their own local clientele. That spreads more happiness than a huge company with a lot of employees.

The fifth advantage of what I call relationship marketing is a term I'll call balance. It helps to equalize the relationship between producer and consumer. Certainly the producer must cater to the consumer; after all, the first rule of business is that the customer is always right. But in today's climate of liability, suspicion and selfishness, it is nice to be able to close the door on someone's face when that person is not a good patron.

Every public establishment has that list of people they call "troublemakers." These are the customers who cause store personnel to roll their eyes at each other, silently communicating the message, "Oh boy, not *HIM* again."

The fact is that some sales can actually cost the establishment money, for a negative gross margin. One of the best experiences I've had to illustrate this point was a woman who bought some beef from us and complained that it was so tough she finally had to put it in the blender, but it was still too tough to eat. Other customers who had gotten other quarters from the same animal said it was wonderful. In fact, one customer was in the process of enjoying her grilled steaks even as we talked on the telephone. This complaining customer loved our chickens, though, and when she came to pick them up I carried the cooler out to her car.

When she opened the trunk, it was half full of cases of Dr. Pepper. Being organic farmers, of course, we generally don't drink soda, and she knew this. She was embarrassed at what I saw, and immediately proceeded to offer the following explanation in her whiny voice: "I got some Dr. Pepper from the store last week and it was a little flat. I took it back and complained and they gave me all this extra to help compensate." She is the same person who looks at sweet corn and squeezes half the kernels before selecting the ears ... of corn she didn't squeeze. She also complained about the chickens. It was some minor detail.

The point is that she wasn't good for business. She was a liability customer. We knew our beef was good, our chicken was good, and our vegetables were good. We stood our ground and just got rid of her. This helps to balance the producer-consumer relationship, so that we concentrate our efforts on profitable sales, appreciative customers, people who "get with the program."

Dealing only with people who share mutual appreciation is far more enjoyable than wondering what egghead is gong to come through the door next.

This is not to say that we never make mistakes. We certainly have, and do. We've refunded money, given free merchandise, and all the conventional things that businesses do to keep customers. But we only do that when we are in error, not when there is some unfounded allegation. And we are free to concentrate our efforts on moneymaking folks. The others get deleted from our customer file.

All in all, the advantages of consumer education and loyalty, product quality, farmer lifestyle and sales balances make relationship marketing superior, in my view at least, to empire building. Life is too short, the family too precious, the farm too enjoyable, to sacrifice those things I hold dear for the mirage of empire grandeur.

Chapter 29

Advertising

Pastured poultry, processed at the farm, is an advertiser's dream. Most advertising attempts to convince, through catch phrases and manipulation of words or figures, that one product is better than another. But when a product stands alone, and far outshines the competition, then advertising is easy because satisfied customers begin spreading the word. Pastured poultry sells itself.

We do not have an advertising budget. But that is not to say, especially in the early days, that we did not advertise. Because pastured, homegrown poultry is so unconventional, I was afraid of being misunderstood and misrepresenting something to people through the confines of conventional advertising. Buying regular advertising on radio or in the newspaper will yield the worst customers. These are the folks looking for a deal. You are not interested in folks looking for a deal. You want folks who appreciate high quality food at a fair price. If you build your business around people who come to you because they have met you personally or were recommended to you personally, you begin immediately to reduce the nameless, faceless weak loyalty engendered by advertising that assumes a large gap exists between consumer and producer.

Early on, I put together a slide program about our farm that I could present to civic clubs, school groups, garden clubs and any group that needed a program. These community organizations are starved for good, interesting programs for their monthly meetings. And Americans do not trust their food. In a normal community, by the time the organization hears about the fire department's activities, the school superintendent's plans, the city manager's or county administrator's problems and the United Way chairman's report, there just may not be much more to hear about.

By offering a program devoted to alternative agriculture and tailoring it to the group's primary interest, I could defend our concepts in person, clarify misunderstandings, and present the whole picture. Over the years, I've spoken to Ruritans, Kiwanians, Rotarians, women's clubs, exchange clubs, FFA chapters, science classes, young farmer chapters, garden clubs and everything in between. The program has been so well received that I'm on my second time around at some organizations. People in these organizations are starved for a really new, workable idea, and respond beautifully to someone who enthusiastically presents passionate beliefs.

Many people have a stereotypical view of an organic farmer as a hippie with a ponytail and sandals on his feet. An unkempt appearance makes people immediately envision unkempt food, and that kills sales. Instead, I cut my hair short, dress appropriately (coat and tie usually) and try to present a professional appearance. This is all part of advertising, and is one reason why conventional advertising doesn't work: it will not break through that hippie stereotype of the organic farmer.

The slide program is primarily educational. It is important not to walk in waving your products and order blanks. Essentially you want to teach consumers about a better way to farm from both an environmental and economic standpoint, and you want to hook them into believing this is the way it ought to be done. Then right at the end, almost as an aside, you mention that they can have these good products by patronizing your farm products.

Such a program requires much research, because the naysayers are always there and ready to debate. This approach is not for the faint-hearted and timid. But probably neither is direct marketing.

If it's appropriate, I'll take a cooked chicken, show how little fat is in the pan, dice it up and provide toothpicks for people to have a sample.

Early on, if we had a customer recommend us to another party who in turn became a customer, we would reward the recommending customer with a dozen eggs or free chicken on her next visit to the farm. Businesses show such little genuine gratitude that it is amazing how little tangible evidence of appreciation it takes to stimulate testimonial advertising and good will.

Because we are doing something so unusual, we were besieged, and still are, by different media for human interest stories. Those are far better than all the advertising money can buy. On a TV talk show, I took a sirloin steak. The host and I concluded the program by eating it on camera, amid his exclamations about how wonderful it was. Money cannot buy that.

And through it all, we've paid close attention to maintaining the high quality of our chicken. The temptation of every growing business is to let quality lag during the expansion. Slippage occurs in control, efficiency and customer relations so that the small business loses its uniqueness as it grows. Through the years, our customers have commented that each year the birds are as good or better than they were the previous year, and that is a praise not taken lightly. We do not ever want to grow beyond our ability to produce the type of chicken our customers have become accustomed to eating.

In the end, the best advertising is customer satisfaction. Chicken produced and processed using our methods will accomplish just that.

Be quick to give things away. At an AARP meeting recently, where door prizes are routine, I took 60 dozen eggs (it was in the Spring when we are invariably overrun with eggs) and gave away 30 of them, one dozen at a time, for door prizes. I was the hit of the day. The folks who didn't get one bought one, so I ended up getting rid of all the eggs and adding some permanent customers in the process.

One of our customers sent a nice note explaining that she just would not be able to get chickens because of a divorce, time and other things. At the end of the season, we took her a dozen birds to let her know we appreciated her difficult situation. Needless to say, she was in tears and her gratitude was inexpressible. She will never forget that, and neither will her friends.

The best way to advertise is to meet the needs of people. That is the real bottom line in your business. If you are truly meeting the needs of people, in a caring, compassionate, honest way, they will respond in kind. Life and business are far more important than money. They are far more important than prestige. I cannot tell you how exciting it is to focus our attention on meeting the needs of people.

Chapter 30

Liability

The direct marketing approach always conjures up fears of disgruntled customers, antagonism and liability. Farmers marketing through conventional channels are somewhat insulated from liability because either the trail from producer to consumer is hard to follow, or, as in the case of broilers, the producer doesn't even own the birds.

The entity that places its label on the final product normally has attorneys on retainer, carries protective insurance, and is careful to join the business and political alignments that help sway public opinion and judges. But a direct marketing farmer cannot afford those luxuries, so what is he to do? And he doesn't employ enough people to be important to the local economy.

First of all, liability insurance is available, and we carry some, primarily because of the number of customers who frequent the farm to buy chicken. If someone would allow their child to run into the field where the bull is, for example, and the child were injured, the parent could sue for damages. That kind of protection is fairly inexpensive.

But what if someone claims that they got sick because of eating our chicken? First, let it be clear that we believe God has led us into this enterprise. If it is to be terminated through no fault of our own, so be it. We are not married to it and it is not the be all and end all of our lives. We refuse to fret and worry about something as remote as a suit. Anyone can be sued at any time for practically anything. No one is truly protected from liability. If we refused to leap any time we were afraid of crossing a stream safely, we would probably never try the first jump. We must do the best we can and leave the rest up to God.

Secondly, we make sure there is no way we can be accused of being negligent. We don't sell inferior products, even at half price. The risk is too great. We keep everything clean and above board, and sell every item as good or better than it was promised. This forces any accuser to invent the reason for his dissatisfaction with our product.

Third, the best protection is a loyal clientele that protects our interests. For example, we've had a customer begin to tell someone about our chicken, and then stop, sensing that the potential customer would not be good for us. Perhaps he is a chronic complainer, or a surly sort of fellow. Money cannot buy that sort of protective loyalty. We do not have to sell to anybody. That allows us to discriminate against bad customers, in effect closing the door on them.

Our customers, more than anything, want us to keep producing these wonderful chickens. Some have offered legal or financial assistance if we ever get into a legal battle. They sometimes ask if we need to borrow some money from them. It is hard for anyone to imagine how loyal customers can be when the link between producer and consumer is as close as ours.

Because we do not purchase media advertising, most people come to us strictly by word of mouth. That means in addition to knowing us, many customers know each other. The relationship bridge extends many ways.

For example, we had a customer who called saying our chickens were no good. She said they smelled and tasted fishy. Through a series of mishaps on that hot, hot day, she had not refrigerated her chickens for nearly half a day. No ice, no refrigeration. I explained what had happened - they had spoiled. But she was adamant and demanded her money back and threatened to have these spoiled chickens cultured at the local medical university. I encouraged her to do so, just to prove what it was. This also called her bluff, proving that I had nothing to hide.

The customer who had introduced her to us, upon hearing about the problem, gave the lady one of his chickens from the same day, which had been properly refrigerated. That was the end of the problem. He did it of his own volition, not only to protect us, but also to prove to her that his recommendation was justified. No one wants to recommend something that turns out sour. Word of mouth advertising is a tremendous protection.

Sometimes a mistake is made, either in calculating weight or price, and a refund is in order. In fact, it doesn't hurt to offer a free chicken next time around. The point is to go overboard in being honest, open, owning up to mistakes, and developing that trust so crucial to customer loyalty.

Finally, we protect ourselves through the

customer file and the order blank. If we have a customer who is not good for us, we just purge him from the file. He simply can't buy our chickens.

That's one of the nice aspects of not operating a public business, where those "pain" customers must be accommodated no matter what. We just do not give them an opportunity to order. We have even pulled customers from the file for being chronically late to pick up their chickens. Life is too short to put ourselves through what we would have to do to please every single person on Earth. Some people simply can't be pleased. Those people can buy their chicken someplace else. It won't be worth eating, but it is probably about what they deserve.

If we see a bad attitude developing in a customer, or we have an unresolved conflict through no fault of our own, that customer simply does not receive an order blank the following year. They forget who we are. That way we protect our interests without having to be nasty or to go through a "showdown." And we can nip a potential problem in the bud before it escalates into litigation or threats.

Certainly for every negative aspect of direct marketing this way, with its incumbent hassles, there is a positive aspect equally rewarding. Instead of looking on all the potential gloom and doom, we prefer to do the best we possibly can, and enjoy the ministry with which we've been endowed.

Chapter 31

Is It Organic?

Organic. Biological. Ecological. Natural. None of the buzzwords is comprehensive. Especially with regard to animals, too many variables in the production and processing procedures exist to cover all the bases with a single cute word.

In the purest sense of the word, our broilers are not organic, and in fact we do not call them that. We call them homegrown, and I would argue that that is superior to being organic.

A broiler can be fed certified organic feed in a confinement house, without fresh air and sunshine, without green salads, trucked for hours to a processing plant that electrocutes the bird and spills feces all over the carcass during evisceration, and be labeled "certified organic." The only poultry prohibition is "battery cages." Rules are liberally sprinkled with subjective phrases like "suitable to the species." Such ambiguous language purposely allows many conventional practices. In animal production, organic describes primarily diet, and everything else is either not mentioned at all or is secondary. I do not believe diet is the primary concern.

Let's suppose that we are going to become purists and eat a 100 percent organic diet. We find a good source of the best organic head lettuce money can buy, and we begin eating. All we eat is 100 percent pure organic lettuce. Anyone would agree that such an organic diet is insufficient to produce health and vigor. Perhaps that is extreme.

Okay, let's suppose that instead of just eating organic head lettuce, we balance our diet with other foods rich in proteins, minerals and vitamins. But we stop washing our hands, we do not launder our clothes, we refuse to sweep the house or wash the dishes, and we even decide to quit taking baths. How long would we remain healthy?

These examples could go on for infinity, but the point is that diet is not the sole, or perhaps even the primary, contributor to health and vigor. The link between our emotional outlook and our physical well-being is well established. A person with a positive outlook on life, a joyous and vivacious disposition, can undoubtedly be healthier eating at McDonald's than a defeated, paranoid person eating at the best organic foods restaurant in town.

It is far more important to have a healthy lifestyle and disposition than it is to worship at the altar of the health food store. Certainly we should try to eat well, too, but my point is that too often, especially in organic agriculture, we focus all the attention on the animal's diet and miss the bigger picture.

The bigger picture includes sanitation, stress, sunshine and general comfort, freedom for

240

behavioral expression and movement, green material in the diet, and finally, but just as important, the entire marketing and processing scenario. The key is balance.

I am constantly amazed that people in Virginia pay exorbitant prices for certified organic broilers flown in refrigerated air freight from California, birds that do not receive green material and are raised basically in a conventional confinement factory house. From a world-view standpoint, it would probably be better for the environment to buy locally produced conventional chicken than to encourage the use of jet fuel and heavy metal to transport that chicken across the country. In the name of one cause, we sacrifice another equally worthy cause.

In past years we have processed true blue organic broilers and found them to be full of tumors, to have such brittle bones that they could not be eviscerated without breaking ribs, and to have enlarged gizzards and look in every way more like a quail than a chicken. The reasons vary from unbalanced rations to typical clay yards for range. But the point is that organic feed is only a fraction of what is necessary to produce a truly dynamic bird. Certainly truly organic chickens are not necessarily bad. But neither are they necessarily good. The term is simply noncomprehensive.

The most important things are to maintain the highest degree of sanitation, ensure an adequate supply of green pasture, sunshine and air, break down the flock into more naturally numbered groups, reduce stress and process in the cleanest, most efficient way. After all these things are done, diet refinements can be made. Otherwise, we get the proverbial cart before the horse. In short, the production and processing models are 80 percent of the bird's quality, and diet is only 20 percent.

The most important aspect of diet is that it be balanced. The birds must have the right ratio of calories to protein, vitamins to minerals, and so on. After that, grain sources can be adjusted.

I certainly do not discourage organic grain production. We would love to use organic grain in our ration. But it is extremely expensive, has to be trucked from out of state, and would force us into the grain storage, grinding and mixing business. The capital required for all of that would dwarf the entire poultry enterprise, and it would require us to greatly increase our prices, which would reduce our potential market base.

One customer, jesting about these ideas, told us our chickens tasted so good he didn't care if we fed them diesel fuel. He said he'd buy them anyway. It was meant for a joke, but it made a sharp point. The difference between organic corn and what we buy at the local feed mill is not enough to show up in the taste or quality of the chicken. The fact is that we have many customers who have switched over to our chicken from those organic birds because ours are superior in every way. They have less saturated fat. The meat is more tender. They have the bloom appearance of health and vigor, and they taste much better. The proof is in the eating.

Some people suggest that we should grow our own grain. The problem with that is that we are not in the grain growing business. Growing a little grain is like being a little pregnant. You either are or you aren't. Just to plant an acre, you have equipment costs that would become the tail that wags the dog on the poultry enterprise.

You would have far more money tied up in grain production than in the poultry production.

Furthermore, our land is not conducive to growing grain. The soil on our farm is thin, the fields are hilly, and the flat fields are prone to surface water flooding. We would exceed the "expectation of the land," as Wes Jackson calls it, to produce grain. Grain production requires tillage equipment, planting and harvesting equipment, storage, handling and drying facilities.

I would much rather patronize someone who is already tooled up for the job to produce the grain and focus my attention on producing the chicken. One of the beauties of value-added rural enterprises is that the dollars generated are high enough to support many ancillary spinoff enterprises. I have my hands full learning about producing the world's best chicken. There is nothing demeaning about letting someone else be the expert on producing the world's best corn. In that regard, we are not competitive, but rather complementary.

We have tried unsuccessfully to get neighbors who grow grain to grow organic grain for us. We guarantee them a market, and a premium price. Have you ever tried to convince someone who didn't believe he could grow anything organically, to grow something that way? It just doesn't work. They have every excuse in the book. The extension agent tells them it won't work; the fertilizer salesman tells them it won't work, and their neighbors and relatives ridicule the idea. And then, supposing they do agree to designate an area to grow the organic corn for you, if they get into midseason and bugs come in, and you're not around, who will keep them from running out there with a little pesticide to salvage their "premium" crop? After all, they don't believe there's any difference anyway.

And it's not just eliminating some things, but getting the soil right. Are they going to learn how to compost just to fertilize a few acres for your special grain? And suppose harvest comes, and they come up short on what they said they would grow for you? How do you know they won't add a couple rows of conventionally-grown grain to the bin to fill your order? After all, they know there's no difference anyway.

The fact is that, unless you can find someone of similar philosophy, integrity and practice, you can't do any better than patronizing the local mill that buys most of its grain from local farmers anyway. In fact, in our area most local grain farmers aren't set up to store grain either: they harvest and deliver it immediately to the local mill for drying and storage. When we buy grain from the local mill, therefore, we often get back our neighbors' production anyway. If the National Organic Standards Board succeeds in requiring 100 percent organic feed in all livestock production, it will either eliminate almost everyone from even attempting to grow certified organic livestock, or it will make liars out of many otherwise honest folks. It will certainly make the prices so high that middle class folks can't afford them. And then we in alternative agriculture will have shot ourselves in the proverbial foot once again. After all, certification depends on the honesty and integrity of growers, just like the Internal Revenue Service depends on the honesty and integrity of taxpayers.

The bottom line is that while we would love to use organic grain - and believe we would if the opportunity ever arose - it is a logistical nightmare. The best we can do currently is to patronize our local farmers through the local grain mill and let the poultry revolution we'd like to instigate generate a spinoff revolution in the local grain growing community. As the old paradigms begin falling, all sorts of new opportunities evolve.

Of course, if you are already growing grain, I would encourage you to use it for the chickens. In fact, this is a way to value add to your corn or soybean crop. Rather than selling it at the local elevator for a wholesale price that scarcely covers production costs, store it and put it through chickens. If I were a cash grain producer, I would jump at the chance to market my grain through animals.

You could keep the manure from the animals, make far more money, and get out of the vicious cycle of producing more and more bushels of corn to make up for the shrinking profit margin. You could plant far less acreage and make more money. The grain required by 500 birds can be grown on about an acre, and that is exactly the amount of land we cover with the chickens. A pasture salad rotation for the chicken pens could be put into your crop rotation. Follow it with corn and then beans before going back into a pasture salad. Imagine being able to plant 20 or 30 acres of grain and net $30,000. If you are already tooled up for grain production, your feed costs could be cut dramatically, increasing your gross margin per bird. Of course, you could sell your heavy metal equipment for cash to get started with chickens. Most of your neighbors will continue producing cheap grain for you.

Perhaps you would want to experiment with some different rations if you are growing other row crops. For example, Chris Wieck in Texas is in the dryland row crop area and uses sunflower seeds and milo as his basic ration ingredients. That was what he was used to producing anyway, so he had no cost to putting it through chickens instead of carting it all down to the elevator. Our feed costs average $1.50 per bird; his average only 70 cents.

The pastured poultry opportunity certainly offers hope for all sorts of people, but for cash grain farmers, it may hold the greatest promise.

The hoopla over organic certification has certainly siphoned off much time and energy from the alternative agriculture movement that could have been better spent in other ways. Remember that certification is primarily a labeling law, and if your product has no label you can call it whatever you want.

Of what benefit is certification if the product is marketed conventionally? The food can be transported far away from the production/processing area and mass marketed. That idea flies in the face of what most of us in the organic foods movement see as the real answer to food quality and sustainability: locally grown and marketed. Regional food sufficiency is the ideal.

Certification allows the same misunderstandings, the same barriers, the same distance, to exist between producer and consumer. That certainly doesn't do our society any good. It certainly doesn't help the farmer.

Being a pass/fail system, it pushes producers toward a minimum standard, instead of consistently pushing them toward a higher standard. Pass/fail courses in school never engender the best performance because there is no extra reward for achievement. This system does not reward any incremental step toward an ideal, and thereby reduces initiatives to excel and innovate better ways of doing things. You're either in or you're out. That is neither charitable nor reasonable.

The fact is that while we may not use organic grain, our other methods far exceed production techniques around the country. The best thing to do is to just opt out of the certification movement. Do a good job and get your own customers through your own integrity. I find it amazing that certification, which is predicated on mistrust of the organic farmer, is only as good as the honesty of the organic foods industry. In other words, it assumes that we need some standard to make sure this food is what its marketers claim that it is because they are not to be trusted. But it relies completely on the honesty of the paper trail, the integrity of the testimony of the people involved. That is illogical.

Certification would raise our price by at least 50 percent because we would be buying grain at double the price, would have to invest in storage and handling equipment, and pay several hundred dollars a year for the certification service. It would not reward us for being any better than anyone else, and it would push our price through the sky, like that of the other organic poultry producers who sell convention-ally.

What good would it do to have a perfect bird at an unaffordable price? I would rather have the best bird at an affordable price, and let perfection come in the Millennium. We are quite content, therefore, to produce the world's best homegrown chicken, and we will gladly compare it to a certified organic bird. Far more can be accomplished with the production/processing models than can be accomplished by shifting from conventional corn to organic corn. That is a refinement that can come AFTER the other weak links have been addressed. We do not apologize because we are not organic. Of course, many people call our birds organic. I don't feel compelled to correct them. If that's the word they want to use, that's fine with me.

The fact is that we and our customers have compared our homegrown birds to those certified dudes, and I'll take them on any day. We always beat them hands down because we have attacked the elements that make the biggest differences.

POSSIBILITIES

Chapter 32

Vision

What if all broilers were produced like they are at Polyface, Inc.? What would the poultry industry look like?

It is hard to answer that question for poultry only, for if such an agricultural revolution did occur, it would reach far beyond poultry. But if we can break out of poultry for the sake of discussion, and zero in on it, perhaps we can appreciate the kind of positive changes that could occur in our culture were all the chickens fit to eat.

First of all, I have no animosity toward the current poultry giants. They are sincerely doing what they think is good business and are merely ignorant of the better way. It would be wonderful for them to begin raising broilers that are fit to eat. But I am not optimistic about that happening.

I envision thousands and thousands of enterprising young farm families around the country beginning to raise pastured broilers. It would take as many as 20 producers to raise as many broilers as one average growout operator. Truly it would take many thousands of producers to meet the demand for broilers in this country.

For many of these families, it would be the first time their acreage ever produced a profit. They would love their land and their animals. They would make good money, be out of debt, and be respected members of their communities. They would enjoy the freedom of working for themselves rather than for the banker or some vertically integrated enterprise. Many could leave off-farm jobs and enjoy the peace and satisfaction of being full-time farmers.

The infrastructure which small scale poultry production currently supports would become mainline poultry agribusiness. Tottering family-owned hatcheries would receive an infusion of capital to upgrade facilities and become more efficient. Manufacturers of small-scale processing equipment would increase in number and size. Many of the big manufacturers would change their inventory, scale down their machines, and use their expert engineering staff to build better machines.

Not a single poultry industry job would be lost. Many maintenance personnel would be needed to install and troubleshoot for the small farm processing operations. Ideally, farmers would cooperate in purchasing processing equipment, mount everything on a trailer and processing crews could go from farm to farm, on a routine schedule, processing that farm's broilers on that certain day.

Farmers isolated from markets perhaps would co-op sales and need experienced salesmen to reach more distant markets. Hundreds of processing plant workers would process their own poultry on their own homesteads and develop fulfilling relationships with their own customers.

Inspectors could join chipping crews, generating biomass for composting offal on the farms. Many inspectors no doubt would enjoy working for themselves, and begin raising broilers fit to eat.

Rural economies would rebound with the increased business activities surrounding the pastured poultry industry. No more would the money go into large processing hubs. No more would so many farmers need to work in cities to earn a living.

Instead, the cities would patronize the counties. Urbanites would carpool, or go for a day in the country, as they patronized locally produced broilers. Children could pet the chicks and learn where their food comes from, instead of being barred from confinement houses. A new mutual respect and admiration between the city dweller and country dweller would emerge, as each became loyal to the other.

Consumers would build relationships with their producer counterparts. A new love and appreciation between the city and country would stimulate cooperative working relationships. Perhaps instead of going to Disneyland for vacation, an urban family would come out to their poultry producing family farm and help build a shed or dig a water line. City children would grow up with ties to one or several farms; these ties would be carried for the rest of their lives for investment, recreational, food and emotional opportunities. The luxury of educating consumers about agriculture would give farmers renewed optimism and completely revolutionize farm family socializing. Instead of "them" and "us," it would just be "us," a mutually-beneficial community of team players who realized that each needs the other.

Many farmers, not satisfied with poultry only, would cooperate with neighbors and begin offering other items like crafts, baked goods, vegetables and other meats. Farm stores would spring up all over rural America as consumers learned the value and joy of buying locally and supporting regional economies. Local economies would grow as dollars stayed in the community. Industrial hubs would decline as rural opportunities increased.

Delighted with the taste of good food, consumers would begin asking for similar qualities in the rest of their food.

Spinoff community businesses would revitalize rural economies. The effect would be much like the injection of money into the Amish community centered around the cheesemaking plant. One cheesemaking plant provides the employment for about 90 families. That includes several dairy farmers, people who raise replacement dairy heifers, farmers who do custom hay making, barn builders, harness makers, blacksmiths, cabinet makers and equipment mechanics.

In the pastured poultry case, spinoff businesses would be cottage industries making pot-pies out of chicken and organic vegetables, a barbecue expert, a hatchery, a processing crew and an expert pen builder. The possibilities are endless. Home-based businesses, which are the wave of the future, could further process and value add the chicken to meet the demand of today's prepared-food fanatic.

Studies show that right now, only 25 percent of chickens are sold as a whole carcass. About 75 percent are further processed into TV dinners, pot pies, fast food chicken and the like. At current trends, by the year 2000, 50 percent of all meals will be eaten away from home. The opportunity for meeting the demand for high quality prepared food is limitless.

Alternative marketing schemes could circumvent inspection tyranny until public pressure forced changes for low-volume businesses. New appropriate inspection requirements for small-scale food processing and handling would be implemented that would more reasonably fit the cottage industry. Indeed exemptions to stimulate regional and entrepreneurial small-scale food enterprises would sweep the country.

The demise of the chemical cartel would be precipitous. Many of those employees would return to the land, either as owners and producers or as workers in large scale composting programs and fertility management.

Hospitals would be mere skeletons, and the entire medical industry would be scrambling to learn about homeopathy, nutrition and wellness. Facilities would still be needed to treat farmers who cut themselves processing poultry. Accidents will always happen.

And what of the myriad obsolete factory poultry houses scattered around the country? They would be used to store hay and bedding material for wintered cattle. Livestock numbers would increase in order to eat all the grass produced by pastured poultry litter. By utilizing all the nitrogen from the manure, forage production would double nationwide. Rather than being lost into the groundwater and killing babies through sudden infant death syndrome, this fertilizer would be utilized by management-intensive beef and dairy producers.

These obsolete houses would no doubt be used

as equipment sheds, or huge composting sheds. By the time they were all empty, the dynamic rural economies would certainly find beneficial uses for them. They could be turned into condominiums. Who knows all the wonderful possibilities?

This may all sound like science fiction, and I do not pretend to think it will actually happen. But it can happen in the hearts and lives of individuals who dare to be different. The opportunity is there for those who will dare to think unconventionally. Such revolutions happen one person at a time.

Chapter 33

Laying Hens: Three Options

I see egg production as a real possibility, and have marketed eggs since I was 10 years old.

In fact, that is what we initially produced in the broiler pens. In the summer I would run the pens outside with the layers in them, and in the winter I would put the pens and hens in the barn.

Feed consumption dropped 30 percent when the layers went out onto pasture, with no drop in egg production. Clearly it was an economical alternative and no doubt produced a much healthier egg. Remember that the saturated fat and cholesterol linkage to high calorie/sedentary production models holds true for eggs as well as meat. Good eggs are a detoxicant, and in fact some alternative doctors prescribe range eggs to patients who need to reduce their cholesterol.

The pens can easily be modified with six foot nest boxes hanging on the enclosed sides to accommodate six partitions on a side, each being 12 inches long, high and wide. A front board about 3 to 4 inches needs to be put on to keep the birds from scratching bedding out. Top access from outside the pen, using a hinged plywood top on the box, facilitates egg gathering. Hanging like saddlebags off the side of the pen, the boxes do not add an appreciable amount of weight and balance each other for moving ease.

The advantage of eggs is that they are not as heavily regulated or as perishable as meat. We've had eggs keep for more than two months in the basement. The main enemy of a stored egg is low humidity.

An eggshell is porous. Eggs are primarily graded on the size of the air cell inside the egg; as the egg ages, the liquid evaporates and the air cell gets bigger. A candler is a strong light that illuminates the inside of the egg so you can see the size of the air cell.

Refrigerators have an extremely low humidity, and therefore should not be the preferred long-term storage facility for eggs. It is actually better to put them in a clean, moist place like a cellar than to put them in a refrigerator. An old design for a makeshift refrigerator is a wooden framework with a shallow pan on top. Burlap hangs down over the sides of the framework, with the ends submerged in water in the flat pan. The burlap wicks the water out and down and the cooling effect of evaporation keeps the eggs, stored on trays inside the wooden framework, cool. Certainly eggs are not as perishable and can be transported more easily than the poultry.

One large drawback with eggs is their seasonal production cycle. Remember that seasonal production is the cheapest way to produce anything. Off-season production is far more expensive and takes its toll on the animal, which is programmed to take a vacation. Unless you use artificial lighting stimulation, you will experience overproduction in the spring and under-production in the winter. Customers must adjust to this, realizing that in the winter they must either curtail egg consumption or buy them out of the store. We give away spring abundance to neighbors, who reciprocate with free cattle hauling, equipment loans and other neat trades. We've never given anything away that failed to multiply before coming back.

One technique to counteract the production problem is to time pullets to start laying in the fall. You sacrifice some production in the winter, but it does not spike as drastically in the spring. It tends to even out the production cycle, although it does tend to sacrifice some in overall volume.

The broiler pens, then, can certainly be retrofitted for egg layers and provide a healthy income. I do not believe it is as lucrative as the broilers, and it is not as seasonal since the layers are a year-round proposition, but the pens will function well in this capacity. In colder areas, certainly the birds would need to be housed more tightly in the winter. A net profit of $10-12 per year per hen is a reasonable expectation.

Certainly all the rules applying to the broilers need to be applied for raising the chicks up to the laying age (usually around 5 months). We use the same ration for the growing phase. But once the birds reach about 16 weeks, we shift them over to whole grains. Whole grains keep them from getting parasites and greatly reduce water consumption.

We prefer old standard American varieties to the hyped-up egg machines bred exclusively for confinement egg production. Our favorite is the Rhode Island Red. These birds, when non-hybrid, are amazingly aggressive foragers and extremely hardy. Since we do not debeak them, our birds consume great quantities of bugs, grass and rocks. Because of this, we find only 40 or 50 per pen is best. That gives about the same utilization level and impaction level as a pen of 100 Cornish Cross broilers.

The small pens allow the same range benefits, production, control and manure spreading that accompany the broiler model. But two other options exist that I want to explore.

The first is the loose housing option. I do not consider a chicken yard an option, even if it is rotated. If it is big enough not to suffer total defoliation in a couple of years, then it will be defoliated near the house and not utilized at all at the far ends. This over-impaction and under-impaction is both inefficient and ecologically inappropriate. If it is small enough to get good utilization, it will soon turn into the proverbial dirt chicken yard.

I know of no working commercial egg production model utilizing a chicken yard, either rotated or not rotated, that offers fresh pasturage and remains ecologically sound, not to mention hygienically responsible. The disease buildup and impaction are simply too great with poultry. They were not created to pasture intensively for any period of time. In nature, birds range over huge areas, constantly moving to fresh ground.

As a result, I began researching a house model that could afford a degree of fresh pasturage without the negative aspects of poultry yards. We have now built a 720 square foot poultry shed that is the result of preliminary research and the subject of ongoing research. To provide plenty of space, not only for exercise but also for stress considerations, we believe 5 square feet per bird to be a target minimum space allotment. We feed whole corn, oats, meat and bone meal and oyster shells, cafeteria style, and free choice so the birds can pick what they want whenever they want it. Cafeteria style means the ingredients are in separate feeders; free choice means they are always available. Whole grains diminish parsites and are easier to store for long periods.

The chickens receive a bale of hay or fresh grass clippings each day to supply forage and encourage scratching activity. Remember that the green material is necessary to increase the vitamin-mineral content of the egg, and to reduce the cholesterol. It is amazing how much hay they will eat.

What they do not eat, they scratch out onto the floor for bedding. Again, the benefits of deep bedding come into play, except that with the layers, who scratch deeply and aggressively, the aeration can come directly from the animal without my having to do anything. Sprinkling some whole grain on the bedding will encourage deep scratching. As long as I keep enough carbon (hay leavings) going into the system, the birds will aerate the bedding and maintain the composting program by themselves. Of course, other carbon-aceous materials can be added, like straw, leaves, wood shavings and sawdust. Research conducted in the 1940s indicated that if the bedding is at least 12 inches it will supply enough bugs to give the birds all the animal protein supplement they need, eliminating the need for the meat and bone meal. That is an idea we are attempting to verify now.

Of course, the birds must receive grit because they cannot get it from the range. They need healthy amounts in order to grind up the whole grains they receive. Any small rocks thrown in occasionally in a pan will work.

Other early research showed that any time the number of chickens in any one group exceeded 300 members, efficiency dropped. That means if we had a 5000 square foot house for 1000 layers, we would want to divide it into four quadrants of 250 birds apiece. Sheer numbers aggravate stress, even though floor density may remain the same. Remember the illustration about living in a crowded stadium; it's okay for an afternoon, but gets old over time.

I wonder how many disease problems could be solved in confinement poultry houses if this simple rule were followed. Of course, partitions would be too logistically complicated to merit serious consideration from poultry experts. They would much rather find out what drug would keep the birds from dying. That is far more romantic. And it keeps the pharmaceutical firms happy. Wendell Berry points out in *The Gift of Good Land* that what is wrong with us creates more GNP than what is right with us. I think the loose housing alternative does provide a workable, commercial intensive production model for producing high quality eggs. The primary secrets, as I see them, are: plenty of forage, whole grains, composting bedding, plenty of space, and groups no larger than 300 per flock.

The third model, and one that we have used for many years, is the portable *Eggmobile*. This is an extensive grazing model, as opposed to the pens, which use an intensive grazing model.

Several years ago I began pondering how I could eliminate the soil impaction of a chicken yard, reduce feed costs, feed chickens bugs and fresh green material without taking it to them, and give the layers unlimited fresh range. The *Eggmobile* was born.

I built a portable 6 foot x 8 foot chicken house on bicycle wheels and made a yard from 3/8 inch rod welded into rectangular gates, 10 feet long and 4 feet high. The gates, in two sets of three, stood up like an accordion fence, a hexagon off the rolling house, which could be pushed from place to place. Rotating the yard around the mesh-floored house gave several days' fresh range per house-move.

It worked well enough that I abandoned the portable yard for total free range. A 3-point tractor hitch hookup on the house replaced the bicycle wheels. I could totally free range the birds and move them easily over long distances. My object was to see how much feed they would pick up and what they would do to cow pats in the paddocks where cows grazed. The prototype provided enough answers that we expanded our laying flock to 100 birds and built the 20 foot x 12 foot trailer model. (Photos 33-1 and 33-2)

The most dramatic difference between limited range and total free range is in feed consumed. In the broiler pens, the 100 birds were eating about $3.50 to $4.00 (still 30 percent less than typical confinement) worth of feed per day, much of it soybean meal and meat scraps, which are both protein sources and quite expensive. After being placed in the *Eggmobile*, their daily feed cost dropped to about 70 cents. They refused to eat soybean meal and ate very little meat scrap. They continued to eat the whole grains and oyster shell, cafeteria style, which allowed them to

262

Photo 33-1. *The 12 x 20 foot* Eggmobile. *Note the trapdoor and chicken ladder.*

Photo 33-2. *Joel gathers henfruit from the* Eggmobile *nest boxes.*

adjust their intake. Their intake of the different components fluctuates with the season.

Egg production dropped slightly, but instead of a dozen eggs costing 60 cents in feed, it now costs less than 20 cents.

The real benefit, of course, is to the livestock. The hens will range up to 200 yards from the house, ranging the first day fairly close and going out in ever-widening circles for about 3 or 4 days until they reach their limit. As they exhaust the immediate range food supply, they begin eating more grain.

Because of their reduced protein consumption, I figure the 100-bird flock is eating about seven pounds of proteinaceous critters per day, the bulk of which are insects (grasshoppers, crickets). A significant portion of their protein, judging by the way they totally obliterate cow paddies, consists of fly larvae. Cow paddies scuffed up by hoof action, tractor tires or my boots are not scratched by the birds. They pick only those unmolested ones permeated with bug holes.

By moving the chickens a paddock behind the cows, we allow them to take advantage of the newly uncovered insects, and the fly larvae which hatch in four days, and they help spread the droppings. I've concluded that the *Eggmobile* would be worth having even if it were stocked with roosters. As pasture sanitizers, the layers can't be beat. In nature, birds always follow herbivores, cleaning up after them. Cow manure contains the proper balance and type of enzymes for efficient poultry digestion.

Another advantage of this approach is that housing costs become minimal. The house, instead of being living room, kitchen, bath, bedroom and recreation (gymnasium?), is merely bedroom and snack counter. Just one square foot of space per bird appears adequate for comfort. Every morning they can roam fresh range to their hearts' content. This reduces housing size by five times compared to the loose housed method, and by two or three times from the intensive pen-housed.

The nest boxes line the side of the trailer-type house for easy access from outside. A front and back door make it easy for us humans to enter, and a small door and ramp allow the birds easy access. I usually close their door at night for predator control. A hinged board in front of the nest boxes keeps the chickens from roosting in and soiling them. I close it in the evening when gathering eggs.

The house height is 5 feet on the lower side and 7 feet on the upper side, making a single slant shed-type roof. The high side has a 2 foot band of 1 inch poultry netting at the top under the eave for ventilation and one end has 3 feet of translucent screen glass for more light. Otherwise, the walls are covered with steel roofing. For winter, I put down sheets of plywood and throw in a couple bales of hay for bedding. The mesh floor is too cold in the winter. If I were rebuilding it, I would make the roof flat. We can always park on enough of a hill to get the roof slanted.

Sometimes in a free range model hens lay in fencelines and under shrubs or other secluded places. Such laying patterns may be interesting, but daily Easter egg hunts can get time consuming.

We've never had any bird lay an egg outside the *Eggmobile*, and I think the reason is quite simple. The birds are never familiar enough with their surroundings to find alternative nesting places. Furthermore, because the ground is fresh and new, with plentiful food, they are busy finding things to eat, rather than secluded places. Familiarity and extra time encourage the birds to find and use alternative nesting sites.

Predators have not been a big problem. Not only are the chickens close to the cattle, but the frequent moves apparently keep predators leery of attacking something so unfamiliar in their surroundings. We do lose a few per year, but that's part of the price we pay. If we were more religious about dropping their door each evening, we would probably not lose any.

The drawback of this model is the extensive acreage required. Each *Eggmobile* requires about 50 acres, and that simply is not a viable model for large-scale, commercial egg production. The reason it works so profitably for us is that it works in symbiosis with the livestock. For that, it is unparalleled.

We also use the *Eggmobile* to debug the garden. Every two or three weeks we pull it up next to the garden and let the birds range through the vegetables. By allowing access for only a day, or at the most two, we keep vegetable losses to almost nothing. But if we leave them the third day the birds start tearing up plants.

They clearly prefer bugs, not plants. If we just let them "cream" the area, then, they don't molest the plants. At the end of the first or second day we close the door on them and the following morning tow them back out far away to the pasture where the cows are. The garden is alluring to them, and once they've been there they need to be moved at least 500 yards away to keep them from wandering back.

266

The only way we've been able to make chickens and the garden compatible is by allowing this highly controlled, limited access. Another idea I'd like to incorporate is to hang a couple of Japanese beetle traps on each end of the *Eggmobile* with a piece of PVC pipe gong from the trap down to a pan of water. I've read that the chickens learn what that clickety-click sound means as the beetle falls through the pipe and are there waiting for it when it hits the water. Japanese beetles live in sod, and it would be an ideal adjunct to our pasture sanitation to also get the beetles. We never get finished refining. The refinements last a lifetime. Plenty of fresh ideas, new twists, are out there for all of us to do better next year than we did this year. Isn't that exciting?

Certainly a smaller house, more like my prototype, would be fine for home production on a farmette. Our little prototype, at 48 square feet, was big enough for 30 or 40 layers. A little under an acre per day per 100 birds is plenty to keep the feed costs down as low as ours are. Certainly a compromise would be in order if you don't have as much acreage. And once you drop to half a dozen birds, their consumption is slight enough that moving them around on a farm would scarcely be necessary. The bugs – the cream element – replenish themselves routinely. In the summer, especially during a drought when grasshoppers and crickets proliferate, a 4-week rest period is enough to restore the "cream" to the area. In other words, we can move the *Eggmobile* back on an area within a month after the birds completely "debugged" it and there will be plenty of new bugs.

In this model we must manage for that rest period so that the pasture critters can multiply and provide that "cream" for the birds when they next have access to the area. That is something you can monitor by walking through and watching for bugs, watching for what the chickens are eating, and watching how much supplemental grain they ingest. As the foraging and bug hunting get harder, the grain consumption rises. If you are tuned in to the system, you'll soon get a feel for the area required for access and the length of rest in your geographic region and for your particular setup.

This method certainly offers the cheapest and best eggs. Maintenance in moving it is certainly no more than is the extra feeding and forage requirement of the loose housing model.

These three egg production models all have their assets and liabilities. You must evaluate your situation and determine what would work best for you. Certainly egg production is a viable small farming enterprise that offers consistent cash flow, an opportunity for children to be integrally involved and needed in the farming activities, and an ideal direct marketing product. We allow our hens to produce for two years. They go through a molt late in the fall of their first production year. The second year nearly all the eggs are jumbo size, which customers love.

After two years, we find that their production falls below the cost of feeding them. We dress and sell the spent hens as roasting hens, and the demand is extremely high. They are tougher to dress than young birds, and the meat is tougher. But if cooked slowly, the meat has an exceptionally rich taste that customers dearly love. The standby recipe for old hens in this part of the country is chicken and dumplings.

268

Another option is to oven cook the birds in a big roasting pan, picking off the meat and using it as precooked, boned chicken. It is a delicacy, and quite versatile for quick casseroles and chicken salad since it can be frozen until needed.

Be assured that the only way to run a profitable egg enterprise is to cull old and poor layers. As a young teen, I learned that "an egg a day keeps the hatchet away" is a requirement for a profitable laying enterprise.

Chapter 34

Turkeys

Turkey has definitely entered the mainstream of American cuisine. It is no longer just a holiday food, although certainly that is still the time when most turkey is consumed. In fact, more turkey is consumed as turkey franks, turkey roasts and turkey ham than as whole turkey, or even turkey breast.

Not having raised turkeys ourselves, we consulted with a neighbor, Jon Moreshead, who is dovetailing a turkey enterprise onto our chicken business. He has raised turkeys now for three years, and found both similarities with and differences from chickens. He is raising white broad breasted birds.

The greatest asset seems to be more efficient processing and greater forage consumption. Because the processing procedure is essentially the same for a 15 pound bird as it is for a 4 pound bird, the person-hours required per pound of meat obtained are fewer for turkey than for chicken. The picker needs to be built heavier, of course, but the processing is still cheaper and more efficient.

Turkeys are far more aggressive foragers.

In fact, Chris Wieck in Umbarger, TX thinks that turkeys would probably eat up to 70 percent of their ration from the range if they had adequate pasturage. His area is so dry that he does not have the forage per square foot that we have. He says predators are a major problem, and keeps the birds penned up in a yard all day. An hour before dark he lets them out and they range up to a quarter mile away. Then just before dark he goes out and feeds them in the yard, which entices them back in. He rounds them up and makes sure they are all penned up for the night.

Turkeys require the same warm, dry brooder area, but need the temperature about 5 degrees warmer. They are much more fragile early on, and Jon says they need to stay in the brooder up until about 8 weeks. During that brooder time, they readily eat handfuls of fresh forage. They can consume far more green material than chickens.

Turkeys absolutely cannot be pushed outside onto pasture too early. In fact, Jon has learned that even most organic producers feed antibiotics to turkeys for the first 8 weeks. He said that it seems up to 8 weeks, all the birds want to do is die, but after 8 weeks, they can take snow, cold – about anything – and it won't even phase them.

Turkeys are more prone to parasites. The texts recommend a three-year pasture rest, but that does not assume rapid moving like we do. Jon says that even with pens, you should plan at least a 2-year rotation.

Turkeys need higher supplement rates in their ration. Jon has been using our chicken ration, and found it deficient in brewer's yeast, calcium, vitamins D and E especially. The second year, he added hard-boiled eggs for the first two weeks and a water soluble multiple vitamin pack along with cod liver oil to the water. He added some buttermilk in a separate water fountain to control any coccidiosis that might be present.

By the third year, his most successful batch was raised with only the extra brewer's yeast (about 35 pounds/ton), extra kelp and grit without egg. The hard-boiled eggs were expensive and time-consuming. He had a 90 percent survival rate at 8 weeks and the birds performed beautifully.

The next batch, though, he reported a complete wipe-out, probably because due to scheduling problems he received the birds from the hatchery several days after they had hatched. He says it seems that if they stay in the hatchery beyond their hatch day they are given medications and synthetics that set them up for disaster two weeks later if they are not maintained on that medicated regimen. We have found the same thing true, although not quite as dramatically, in chickens. The best thing is to get them right out of the incubator, within a few hours of hatching. Medications suppress the immune system, reducing its ability to function once the drugs are withdrawn.

Jon built pens similar to ours and moved them once a day. He tried one pen about 3 feet high to accommodate the taller birds, but it was too cumbersome to move and service. He dropped back to the 2 foot level and the birds seemed quite content -- just more "subservient," he said. He only put about 30 birds in a pen because at maturity a turkey is equivalent to about four chickens.

Our experience is that the turkeys don't sell as easily as the chicken. The ratio of 20 percent response from chicken consumers has been fairly constant for the three years and the average number per customer is about 3 per year. Most customers want turkey late in the year for Thanksgiving and Christmas. Jon said consumers need to be educated to the fact that turkey makes good eating all year long. The season can last longer because he has found that the turkeys tolerate cold weather and even snow. He is thinking too about offering frozen birds year round rather than just fresh ones.

From a cash flow standpoint, turkeys do not perform as well as chickens because the 16 week growing cycle is twice as long. That means the up-front costs are far higher than they are for the chickens.

All in all, turkeys offer some assets and some liabilities, just like anything else. I do not consider them an exotic. The price per pound should be a little higher, just like it is in the grocery store. The opportunities for profit are certainly there. On his good batches Jon reports an $8 per bird profit margin. It can certainly be a profitable business or sideline to the chickens. Try it.

Chapter 35

Exotics

Poultry certainly includes far more than chicken. Waterfowl like ducks and geese can be raised for both meat and eggs. Duck is a luxury item in any supermarket.

Pheasants, quail and chukars all provide additional poultry opportunities. While I do not claim to be an expert in these exotic species, I have thought about them quite a bit - long enough to turn them down for several reasons.

The real goal of a business is to get customers to come back. To ensure that, you want a commodity that they not only desire, but will also use routinely. How many pheasants does the average family eat in a year?

It is clearly a luxury item. In fact, most people have never, and will never, eat one - even in a whole lifetime. But people routinely eat chicken, even on weeknights. And that is a much firmer foundation for repeat business than a commodity that people buy once in a blue moon. That is not to say that a market doesn't exist for these exotics, but I think it needs to be stated emphatically that this is a small market, and one that is easily saturated. While just a few commercial producers would fully saturate the quail market, it would take many, many producers to saturate the chicken market.

Opportunities exist, to be sure, in these other areas, but the niche is small, and it is also more subject to the whims of the economy. I am not optimistic about the long-term health of the American economy, given the deficit, the ever-increasing tax bite and the regulatory stranglehold on business. Sometime the debt must be paid, and when that occurs, the economy will falter bigtime. But even then, people will still eat chicken.

People will eat chicken long after they quit buying pheasant and quail. When they can't afford roast duck, they will settle for chicken. Perhaps chicken will become the new luxury, but it will survive long after the exotics have come and gone. Even in the 1929 depression, 70 percent of Americans never lost their jobs.

Strictly from a business standpoint, therefore, I have chosen to stay away from the exotics.

But there are other reasons. Pheasant and quail and chukar are wild fowls, needing far more space and more wildness in their production model. They are more fragile to raise, and more prone to health problems and cannibalism. The chicks are more expensive to buy, which increases the up-front risk and investment, and the birds have a more delicate skin. This makes them difficult to process mechanically without tearing the skin.

Their production time, from hatching to slaughter weight, is longer and this of course decreases cash flow and increases housing requirements. Their egg production season is shorter, because the hens do not lay like chickens. The entire production time, therefore, must be compressed. Whatever money is to be made must be made in a shorter time. This further complicates an already busy production period, limited housing capability and processing ability.

Ducks and geese are certainly more aggressive grazers than chicken, but waterfowl especially are hard to process. The feathers, because of the oil, do not pick easily. These difficulties show why these exotics sell for three and four times as much as chicken. There is no free lunch. They are expensive because they are that much harder to produce and process. In my experience, the return per hour can be no more, and may be less, than for producing chicken.

True, the fixed costs as a percentage of gross receipts may be smaller, but the bigger margin is eaten up in extra labor. What we are really selling is our time. If the exotics give us no more for our time, we may as well stay with the tried and true chicken.

I do think exotics would be healthier in a pastured model than they are in the normal concentrated dirt-floor aviaries I've seen just like chickens respond to a clean pasture. But what if a pheasant got out of the pen? Who is going to catch it? You can just run a chicken down and return it to the pen. No problem. But what about a pheasant? I can envision more time spent chasing birds than moving pens. Also, the pheasant pens I've seen are 8 to 10 feet high. I do not know if they could handle just 2 feet or not.

The bottom line is that chicken appeals to the masses, and is a proven performer in both the production end and the processing end.   In my opinion, exotics are something to try for fun, not for a business.   When you get bored with life, perhaps they could spice it up.   But in my opinion, exotics do not hold the promise that chickens do.

If you are near a hunting preserve that offers caged fowl for recreational shooting, that might be a different story.   If you live near a metropolitan area, upscale restaurants may be a nice market.   Everyone has unique opportunities and we need to be creative in finding and capitalizing on them.   I do not want to discourage anyone from finding a lucrative niche in agriculture.   But I believe it would take a special set of circumstances to justify going to the exotics over the tried-and-true chicken.

APPENDICES

Appendix A

Resources

FEED SUPPLEMENTS

Countryside Natural
  Products
P.O. Box 997
1-888-699-7088
(540) 932-8534

Fertrell
P.O. Box 265
Bainbridge, PA   17502
(717) 367-1566

KELP
Bill Wolf
Salem, VA
1-800-464-0417

SONIC DRIED BREWERS YEAST
NPC, Inc.
P.O. Drawer B
Payette, ID 83661
(208) 642-4471 or
(910) 635-5190

SUPPLIES

Nasco Farm and Ranch
901 Janesville Ave.
Fort Atkinson, WI
(414) 563-2446

Nasco West
1524 Princeton Ave.
Modesto, CA   95352
(209) 529-6957

Northern Greenhouse Sales
(204) 327-5540

Premier Fencing
1-800-282-6631

SUPPLIES (continued)

Stromberg's Chicks and Pets Unlimited
Pine River, MN   56474
(218) 543-4223

Who's Who in the Egg and Poultry
   Industry in the U.S. and Canada
Watt Publishing Co.
122 South Wesley Ave.
Mt. Morris, IL   61054
(815) 734-4171

PROCESSING EQUIPMENT

Pickwick Company
1870 McCloud Place, NE
Cedar Rapids, IA   52402
(319) 393-7443

Brower
P.O. Box 2000
Houghton, IA   52631
(319) 469-4141
(319) 469-4402 FAX

Rob Bauman
RD 2, Turner St.
Oxford, NY   13830
(607) 843-7415

Ashley Machine, Inc.
901 N. Carver St.
P.O. Box 2
Greensburg, IN   47240
(812) 663-2180

Who's Who in the Egg
   and Poultry Industry
   in the U.S. and
   Canada (see p. 279)

Jako, Inc.
6003 E. Eales Rd.
Hutchinson, KS   67501
(316) 663-1470

WATERERS

G&M Sales of Virginia
Highway 11 South
Harrisonburg, VA 22801
(540) 433-9156

BASIC H SOAP

Shaklee U.S., Inc.
San Francisco, CA
   94111

HATCHERIES

Clearview Stock Farm
and Hatchery
P.O. Box 399
Gratz, PA 17030
(717) 365-3234

Heatwole Hatchery, Inc.
P.O. Box 271
Harrisonburg, VA 22801
(540) 434-6738

Hubbard Farms
Walpole, NH 03608
(603) 756-3311

Ideal Poultry Breeding
Farms
P.O. Box 591
Cameron, TX 76520-0591
1-800-243-3257

Marti Poultry Farm
P.O. Box 27
Windsor, MO 65360
(816) 647-3156

Mt. Healthy Hatcheries
Inc.
Mt. Healthy, OH 54231
(513) 521-6900

Moyer's Chicks Inc.
266 E. Paletown Rd.
Quakertown, PA 18951
(215) 536-3155

Murray McMurray Hatch-
ery
Box 458
Webster City, IA 50595
(515) 832-3280
1-800-456-3280

Reich Poultry Farms,
Inc.
1625 River Road
Marietta, PA 17547
(717) 426-3411

Ridgeway Hatcheries
615 N. High Street
LaRue, OH 43332
(740) 499-2163

Stromberg's Chick and
Game Birds Unlimited
P.O. Box 400
Pine River, MN 56474
(218) 587-2222

Yoder's Hatchery
Rt. 2, Box 140-B
286 Jarrell Road
Tylertown, MS 39667
(601) 736-1000

PROBIOTICS

Robert Clements
(Conklin Corporation)
Rt. 1, Box 522
Weyers Cave, VA 24886
(540) 234-8212

280

INFORMATION

Periodicals

ACRES USA
P.O. Box 8800
Metairie, LA 70011
(504) 889-2100

American Small Farm
9420 Topanga Canyon
Chatsworth, CA 91311
(818) 727-2236

APPPA Grit!
American Pastured Poul-
try Producers' Asso-
ciation
5207 70th Street
Chippewa Falls,WI 54729
(715) 723-2293

**Countryside and Small
Stock Journal**
N2601 Winter Sports Rd.
1007 Luna Circle NW
Withee, WI 54498
(800) 551-5691

**Holistic Management
Quarterly**
Albuquerque, NM 87102
(505) 842-5252

Permaculture Activist
P.O. Box 1209
Black Mountain, NC
28711
(704) 683-4946

Quit You Like Men
P.O. Box 1050
Ripley, MS 38663-9430
(601) 837-4596

Small Farm Today
3903 W. Ridge Trail Rd.
Clark, MO 65243-9525
(800) 633-2535

Small Farmer's Journal
P.O. Box 1627
Sisters, OR 97759
(503) 549-2064

Stockman Grass Farmer
P.O. Box 2300
Ridgeland, MS 39158
(601) 853-1861
1-800-748-9808

Appendix B

Newsletters

January 1984

Attention friends of Polyface, Inc.:

Greetings from the farm of many faces. We trust your Christmas and New Year's memories are rewarding.

This is our first clientele newsletter. We hope it will be a periodic production to encourage communication and mutual satisfaction.

VEGETABLES: Last summer for the first time we sold our organic garden produce. As a result of several sundry sales of surplus veggies, we learned that people need this high quality produce. To expand this enterprise in 1984, we decided to offer cabbages, peppers and squash in quantity. Only if we have a surplus of other produce will we offer more variety. These three items, however, usually do well enough for us to feel good about promising them. We will order fungicide-free seed at the end of January to grow sets in our greenhouse for later transplanting. We need to know how many of these vegetables you think you might want over the course of the

growing season.   Prices will be whatever
supermarkets are charging at the time.

Cabbages will be ready sometime between May
1 and June 30.   Peppers and squash should be
available from July 1 until frost.

POULTRY: Because of the tremendous response to
the 400 almost-organic fryers we raised in 1983
we are looking at this item closely for 1984. But
frankly, we had problems.   Not with the production, but with the processing. We sold them for
65 cents a pound. After monitoring grocery store
prices for awhile, we discovered radical fluctuations between 59 cents and 85 cents per pound. Not
including our 50 manhours of processing time, we
netted only 50 cents per bird. Honestly, that's
not enough.

Unless we find a way to net more profit we
can't justify the time.   We are examining other
processing options to reduce its time.   We are
also convinced that we would have to charge 85
cents per pound dressed to make the project -
which we enjoyed - worthwhile.   If you would be
willing to pay that price, please call us.

Right now, we aren't sure whether or not we
will raise them. But since we probably won't be
mailing another newsletter prior to our decision,
we would like to know if there is enough interest
to pursue poultry production.   Then if we do
decide to raise them again, we would call you who
are interested. Although we fear boring you with
these details, we would rather level with you
about our problems than to, without explanation,
either raise the price or refuse to raise the
birds.

BEEF: We have three animals ready to process the first of June. Since our sales went well last year, we are changing some procedures. We will offer front quarters, hind quarters and halves at three different prices to conform to the meat trade. Beef commodity analysts disagree on what will happen this year so we aren't quoting prices yet. All we promise is that our prices for better beef will be lower than grocery store prices for smaller quantities.

Several of you have said that your beef isn't lasting like you thought it would. That is understandable with tender, organic, lean beef. If you would like to get another quarter in June to tide you over until October, when we hope to offer beef again, please call us. We will be glad to answer any questions, too.

We regret that because of a problem we had last fall, we must request a $25 downpayment on all beef orders. It is illegal for us to sell it after processing and this will, we believe, help to insure our orders.

So far, we have heard nothing but praise about the beef. If you have not been satisfied, please call us. Sometimes a cooking technique can make a world of difference. You who are satisfied, we appreciate your encouragement. If you are happy enough with the product, please mention us to your friends and neighbors. We hope that we will not have to send any more of our special animals through conventional wholesale-packer-supermarket channels.

RECREATION: A questionnaire that several of you filled out when you purchased beef revealed significant interest in camping, hiking, picnicking and fishing. We appreciate your candor. We are researching these areas and hope to make some positive reports by year's end. Progress comes slowly. Maintenance is what takes our time, isn't it?

FIREWOOD: Lord willing, we will also have plenty of firewood available during the summer and fall. We are cutting now and stockpiling it near the house.

At the risk of sounding condescending, we know how easy it is to delay. Before now, we have marketed our items by telephone and thereby received an immediate response. But we have an expanding customer and product list which preclude using the telephone exclusively. Truthfully, we fear everyone setting this newsletter aside and forgetting about it. Please don't.

As we close this letter, a milestone for us, we again extend heartfelt appreciation for your interest in our pursuits and biological farming methods. If you have never been here to see firsthand what we are doing with our little niche of God's creation, please arrange for a tour. We deeply appreciate God's blessings, viewing our endeavors as a ministry in every sense of the word - to consumers, agriculturalists and our society.

If you have any questions about any of these items, or if you have suggestions, and certainly (ha!) if you have orders, we are as close as your telephone or mailbox.

Expectantly yours,

Joel F. Salatin, Pres.

March 20, 1985

Dear supporters:

Happy Spring! We trust this newsletter finds you anticipating nature's awakening and marveling at the Creator's handiwork.

BEEF: We guarantee that ours hasn't eaten broiler litter, the new rage recipe. Neither is it eating antibiotics, which have been blamed recently for a death traced to a North Dakota feedlot.

The next processing date is early June. The price is the same as last year: front quarters, $1.28 per pound, hind quarters, $1.55 and halves, $1.35. The price includes cutting, wrapping and freezing. Hearts, livers, tongue, kidneys and bones are also available. Please place your order by April 15 since supplies are quite limited.

POULTRY: Because of the tremendous demand for our broilers, we invested in some snazzy processing equipment to accommodate all orders. Depending on demand, we plan three rotations to be ready about June 4, July 30 and Sept. 24.

These birds will weigh 3-4 pounds and the price is $1.00 per pound. We will dress and cool them to well water temperature.

FIREWOOD: Again, a tremendous demand has kept us cleaned out nearly all winter. Effective immediately, the price is $25 per pickup load. We cut it to length and pile it here at the house. You haul and split. Call before you come to make sure we have some down. It is sold on a first come, first served basis.

VEGETABLES: No demand for two years. If you want something, let us know, but we are not planning to plant for market this year.

286

<u>RECREATION:</u>   We now offer on-farm vacations. Your accommodations are in the two-bedroom fully-equipped mobile home adjacent to our yard.   Two meals per day are available.   We would welcome you or your acquaintance to hike, camp, explore, learn about organic food production, talk with the animals and "get away from it all." Rates flex with your meal desires and length of stay.

<u>SLIDE PROGRAM:</u>   We offer this 20-30 minute, fast-moving program to promote conservation, eco-farming, nutrition and healthy living as practiced by Polyface, Inc.

Please accept our deep appreciation for your encouragement and support.   Call us today with your orders, questions, criticisms and to schedule an on-farm tour.

Expectantly yours,

Joel F. Salatin, pres.

Polyface Inc.

Spring 1986

Dear supporters,

Happy Spring! Please allow me to crank your thoughts ahead for a moment as we share together.

For your convenience we are enclosing a simple order form and self-addressed envelope with this newsletter. The order blank has plenty of room for you to write additional desires or clarification. After receiving your order, we always get back in touch with you, so don't worry about misleading us. We just want a rough idea of what you want. We hope this tool encourages responses, enabling us to plan schedules better.

Many of you have already placed orders -- if that is the case, your order blank is marked and all we need is a breakdown on how many when. If you're flexible, indicate that and it will be a big help to us.

Just a word about the ground beef. This is from animals we don't believe measure up to Polyface cuts quality (slower-growing animals or cows being culled from the herd for some reason). Depending on size, quarters may yield anywhere from 30-60 pounds. Those of you who have purchased these animals from us in the past know that the ground beef quality is absolutely superior. And, as with the rest of the beef, it's biologically grown. All prices are based on hanging weight and include cutting, wrapping and freezing.

Because our homegrown broilers are in such demand, we plan to add two batches to the rotation. Last year orders approached 1,000 and we hope to top that figure this year. Remember, you must order your birds before slaughter. The price is the same as last year, $1.00 per pound dressed.

Two other quick items. We hate junk mail, so if you want us to stop sending you this newsletter, tell us and we will gladly oblige. Secondly, feel free to introduce us to neighbors and friends. We give discounts to customers who introduce us to others who become customers.

By all means, communicate with us if you have questions, advice or encouragement. Someone is always here at one of our two phone numbers. Thank you for your patronage. Come and see us.

Joyfully at your service,

Joel F. Salatin, president

Polyface, Inc.

Fall 1986

Dear supporters,

If you've had as good a year as we have, you're on cloud 9. The cows and chickens have never looked better or been more contented. Truly we have much to be thankful for.

Fall beef will be available in early November. Some orders have already come in, so please respond soon while supplies last. One quick pitch. Forage-fed beef has less fat, and the fat that is there is UNSATURATED. Grain-fed beef and especially steroid-fed beef (nearly all commercially-produced beef) contains SATURATED fat due to altered metabolism. Seems like the more we learn, the more confidence we have in our special products.

The demand for our homegrown broilers is continuing to exceed our wildest expectations. We produced about 1200 this summer and have had to turn away many late-comers who asked for some after the last batch had been ordered. We only order chicks to fill the orders we've received, so there's a fairly healthy lag time (12 weeks at least) between your order and the finished product. We apologize to those of you who got a "no" late in the season, but trust you'll understand the situation and place your order early for next year.

By all means, communicate with us if you have any questions, advice or encouragement. Someone is usually here at one of our two phone numbers. Thank you for your patronage. Come and see us.

Joyfully at your service,

Joel F Salatin, president

Polyface, Inc.

Spring 1987

Dear friends,

What an exciting time to live. Never have the challenges been more challenging or the opportunities more stimulating.

We greatly appreciate the many of you who have already ordered broilers for May and June from our fall newsletter. The response was wonderful. If we already have an order from you, we've noted it on the enclosed order blank.

We are now taking orders for broilers that will be ready about July 18, Aug. 11, Sept. 5, Oct. 3 and Nov. 2. We've added the November batch to reduce the time until the May broilers. Remember, we operate on a 12-week lag, so we need orders now to know how many chicks to purchase, especially for July. You still have some time for the later batches.

It's time to order beef for June. Mark the order blank as you desire. Several orders have already come in -- supplies are limited. Because demand for pure ground beef (made from our cull cows) is so high, last fall we purchased two calves to help us meet this market. Although they were not born here, they've been here long enough that we think they are acceptable for ground beef consumption. We want to be totally open about our products and that's why we're telling you this.

We raised a couple of pigs for our household this winter and to our astonishment found that some of you love pork. The pigs tilled the garden for us and had a ball. The meat is very lean but firm and tasty because of all the rooting and plant material.

We've decided to offer pork for the first

time. We will sell it by the half, hanging weight.
We will cure the bacon, shoulders and hams, if you
desire, using Teresa's old family recipe (sugar
cure). All uncured meat would be available in
mid-November. The curing takes about 6 weeks. If
you want it smoked, you'll need to take it
elsewhere to get that done. We're trying to stay
simple initially -- maybe we can get more
sophisticated in the future.

We also plan to add eggs by fall. The hens
free range from a portable house that we move
around with the cows. Vegetables are available
on a surplus basis and in season. Ask if you want
some. We hope to sell eggs and vegetables on
broiler pick-up days at least, and other days as
available or desired.

We sincerely appreciate your encouragement
and patronage. We're looking forward to serving
you what we believe is the best meat in the world.
Enclosed are a couple of business cards for you
to pass along. If you have any questions at all,
call. Come and see us.

Yours,

Joel F. Salatin, president

Polyface, Inc.

SPRING 1987 ORDER FORM

CHICKEN

Indicate how many. Price is
$1.00 per pound dressed.

_____ July   _____ October

_____ August   _____ November

_____ September

Would you like eggs

($1.00 doz.)?

Yes   No

Vegies?   Yes   No

If yes, what kind?

June BEEF

All prices based on per pound,
hanging weight, and include
cut, wrap and freeze.

_____ Front quarter   ($1.28)

_____ Hind quarter   ($1.60)

_____ Half   ($1.38)

_____ Quarter ground ($1.19)

_____ Liver   ($1.60)

_____ Heart   ($1.30)

_____ Tongue   ($1.30)

November PORK

_____ Half   ($1.65)

CLARIFICATIONS OR COMMENTS WELCOME (USE BACK IF NECESSARY).
THANK YOU!

293

Spring 1991

Dear folks,

This is our annual formal communication and we trust that, although it is a little longer than normal, you will take just five minutes to read it thoroughly. We will bare our hearts to you a little, "sound off" a little, explain a little and help you a little. First let us sincerely thank you for your patronage and encouragement. We brag on you everywhere we go.

First, let's discuss transportation. As you know, we do not deliver. We want to encourage carpooling for those of you who want to spread the pick-up burden. If you will enclose a stamped, self addressed envelope with your order form, we will send you names, addresses and telephone numbers of other customers in your general area. We realize some of you enjoy the drive out, looking at the animals and hiking around and do not want to carpool. That is fine. Don't send an SASE and we won't divulge the fact that you're a customer. Send an SASE, and we'll tell all.

Along that line, let us bare our hearts to you for a minute. The Polyface ministry is to produce the best food in the world. That is our focus. That consumes our energy, our time, our dreams and our expertise. Whenever we deviate from that focus, we begin to fail.

That failure shows up in a stressed lifestyle, economic setbacks, and food that may not quite measure up to our standard. We do not believe we are responsible to feed the world, or to build an empire. We enjoy the efficiency of smallness. In the world of living things, bigness has many drawbacks. Cleanliness, lack of stress, timely management and proper care are much harder to maintain on a large production scale. Tragically, many good small businesses become bad big businesses.

We do not intend to hire employees, and begin mass marketing. Nothing is as efficient as a small, diversified, wholistic family farm. We cannot hire the same level of care, hard work, commitment and expertise that we have as a family farm. While we may take apprentices or pass through offers, we do not want to become dependent on hired labor.

In the modern American business community, people get used to buying whatever service they want. Usually enough dollars, held out as the proverbial carrot on the end of the stick, can get movement out of anyone for anything. If a consumer wants a special consideration, a special service, he just offers to pay extra and it is done. But there is more to life than money, and Polyface will not be bought.

What we are driving at here is the whole issue of amenities and size. Bagging, delivering, freezing. These things seem small and simple, but multiply that by more than 350 customers from all over Virginia and out of the state, and then realize that when we process chickens we begin before 5 a.m. and have regular chores to do on top of the processing, and there simply are not enough hours in the day and enough energy in the individuals to do everything anyone may want ... at any price.

Now that is not to say that occasionally a car breaks down or there is an emergency that demands that we pick up the slack for you. All of you know that we have been willing to do that. But we can only take up so much slack. And offering to "pay for it" doesn't make it any easier for us.

In the past, we have generally assisted when a customer wanted his chickens bagged. This year we will have an extra one or two tubs and another table where those who want to bag can work while we go ahead and help the next customer. Sometimes our biggest need of the moment is to sit for a couple of minutes. Please do not think we are lazy. We may need a little breather so we'll have enough energy to take care of the chickens out in the field or fix supper. Our point here is not to complain, but to express our limitations. If we fail to understand our limits, we get cranky because we feel overworked and underpaid. Then the quality of our food becomes compromised because we can't get around to all the things that need to be done exactly when they need to be done.

Some people refuse to even think about driving out to pick up chickens. Yet they will spend hundreds of dollars on a vacation fling for personal enjoyment and eat food that's not fit to eat. We need customers who are willing to meet us halfway. We will gladly produce the world's best food. But the only way we can enjoy producing it at a reasonable price is to stop at the limits of our energy and efficiency. Why does food of inferior quality cost twice as much as ours? Often the answer is amenities. Those refrigerated trucks and teamsters don't come cheap.

Most of you have and continue to be wonderfully supportive about meeting us halfway. You are wonderful. And our partnership will ensure a long and lasting mutually beneficial relationship. To people who want their food bagged, frozen, cooked and delivered, we only offer our apologies. This is a two way street.

Now to a related topic. Price. It has been

said that the cheapest food is not always the least expensive. Certainly that is true. Conventionally produced food generally erodes the soil, pollutes the water, air and environment generally, makes people sick, hurts the rural economy and debases the food supply. What does all that cost? Nobody knows, but it would sure splash red ink all over 49 cent fryers and 79 cent hamburger. The point is that society, including future generations, picks up the tab for today's shortcuts.

People who are used to patronizing health food stores think our prices are extremely low. Others, used to more conventional food sources, think they run high. Regardless, ours is a true price and reflects everything. No hidden societal costs. Our prices are set to pay expenses plus a modest salary. The laborer is worthy of his hire. And the stereotypical "poor, dumb farmer" image is not one we endorse. Why should we not earn the equivalent of our customers' median income? We must receive enough compensation to keep us optimistic and joyous in our efforts. Our prices should not be compared with the local supermarket's. The philosophies and food qualities are as different as apples and oranges. While we promise not to gouge or to jack up the price because it's "natural," we also promise not to be agrarian serfs.

Thank you for bearing with us through this discourse. We've always wanted this newsletter to be more than commodities and business, and trust that this discussion has been helpful. Now let's move to business.

EGGS The eggmobile is in full production and we have plenty of eggs. Free ranged from a portable henhouse, these are happy birds. The eggs will keep up to 3 months in refrigeration because of

297

their potency. Come on out and get 20 dozen.

BEEF This is our building year. We are keeping all of last year's heifers in order to increase the herd. That means we must temporarily reduce the number we sell, and that means we will be quite short this year. We do not know exactly how many we will have, especially the cows that we turn into ground beef. We tentatively plan to have two or three in June and about four to six in the fall. Please order now for the year and we'll assign what we have first come, first served, to previous customers only. We apologize for the delay. Eat more poultry.

CHICKEN The order blank dates are tentative. Last year the hatchery missed one week which threw two of the batches a week off schedule. We normally process Wednesday, Friday and Saturday of that week, which starts on Monday.

STEWING HENS This year we will have some stewing hens from the eggmobile. The old standby for these birds is chicken and dumplings. We like to cook several together and pick off the meat, then freeze it for quick addition to casseroles or chicken pot pie. The meat is more rich in flavor, but tougher than the young birds.

RABBIT Daniel, the enterprising 9 year old around here, is still desperately trying to fill all the rabbit orders that came in last year. We greatly underestimated the number of people who were interested in this most dense of protein meats. He also learned that his price was too low. Those who ordered last year and have not yet received rabbit will get it at the old price. New orders will be honored as rabbits become available.

VEGETABLES We plan to have cabbage, carrots,

298

beets, tomatoes, yellow squash, zucchini squash, butternut and cushaw (pumpkin) squash, peppers, green beans, cucumbers, sweet potatoes and Swiss chard as available.

READY-TO-LAY PULLETS Available in early winter, these traditional American breeds come complete with range experience and a whole beak. Perfect for backyard home production, these pullets begin laying in January and last up to three years.

We were delighted with your support of Jon and Susan Moreshead's turkeys and lamb last year as they dovetailed their pioneer efforts with ours. Please turn this page over and catch up on their thoughts. Then fill out the appropriate order forms, enclose them both in the same self addressed envelope, put on a stamp, and enclose an SASE if you want carpooling. We'll sort it all out when it gets here.

Remember that we invite your visits, questions and criticisms. We are as close as the telephone. We sincerely appreciate your patronage and your loyalty and just can't wait to see you during the year to strengthen our relationship, building bridges between the rural and the urban, between producers and consumers, and betweens families and their future. Thank you for letting us serve you.

Warmly,

The Salatins of Polyface Inc.

Spring 1992

Dear folks,

Welcome to the new production year. It has been a year of unprecedented changes here at Polyface Inc. and we appreciate your being a part of this dynamic enterprise. Since this is our only formal annual communication with you, we'll use this forum to update you on our goings-on.

The first major difference you may notice when you come to the farm next time is a timber cut on the mountain behind the house. As many of you know, our farm extends up and over Little North Mountain, but we've never had access to that 400 acres of forest. Since coming to the farm in 1961, our family has always dreamed of a road "to the top of the mountain." After many years of searching and consulting, we traded about 50 acres of timber in four areas for 3.5 miles of all weather road. We can now harvest mature timber, clean up dead, down and diseased wood, and husband it. It opens up completely new and exciting stewardship possibilities. The cut areas will grow back quickly and the view ... well, you'll just have to see it.

The second major change is a 720 sq. ft. "RAKEN" house on the end of the equipment shed toward the house. What is a RAKEN house? It's a combination rabbit and chicken house. Daniel needed a permanent facility for his rabbit production and we wanted to research loose-housed, deep bedding laying hen production. We combined the two in a symbiotic relationship. The rabbit pens hang at eye level and the chickens are on the floor. The rabbits benefit from the warmth of the chickens' body heat in the winter, as well as the composting deep litter on the floor, and the chickens pick up bits of grain the rabbits drop and keep the bedding stirred, aerated and clean under the rabbit pens. It is perhaps the most soothing place on the farm to just go in and sit. You'll want to step inside when you come.

The third major change is that we just purchased ... uh, well, we hope you'll help us purchase, a brand new four-wheel drive tractor with a front end loader. It finally arrived a few days ago and we are excited about preserving Joel's back now. We've simply outgrown our ability to shovel everything.

Finally, we are nearly finished with the L shed on the far end of the equipment shed, and plan to use it for storing equipment and dimension lumber. The two-storey greenhouse on the south end of the house is now painted and ready for the glass, which we now have in the poultry processing shed. We hope to get it put in before the first batch of chickens.

Isn't all this exciting? And how grateful we are to each of you for playing an integral part in our dream reality. How thankful we are that we have not gone the way of many in the organic foods movement; the way of conventional marketing methods where the producer and consumer do not know each other. Where California producers supply Virginia consumers. How much more enjoyable to build personal relationship bridges, to effect mutual appreciation between rural and urban, between producer and consumer. We trust that you will use our alternative food and our alternative thinking to pursue alternative choices in your lifestyle, entertainment, education and spiritual pursuits. We are in far more than the commodity business.

Last year's letter, you may recall, addressed carpooling. The response was unbelievable; NOBODY was interested. Well, a couple were, but that's all. We don't intend to address the issue again. It's dead. Thanks for coming.

POULTRY

On the order blank, we have articulated our tentative processing dates. We normally process Tuesday, Wednesday, Friday and Saturday. Remember that we call you a week in advance to get the exact day you'd like to get your chickens, but we thought it might be helpful to put it all down in black and white. Remember that we process in the morning and you come between 1 p.m. and 5 p.m. on that day. The best way to get the birds is in a cooler. You need about 2 quarts of cooler space per bird. If it is a warm day, we suggest you bring some ice. Poultry is extremely perishable. We buy bags in 1,000-count lots, so you can get them cheap from us if you'd like. You are free to bag them here; we provide a separate table for you to do that at your leisure.

The broilers, of course, are 8 weeks old, range raised in floorless pens which we move daily to fresh pasture. Our ration consists of corn, peanut meal, soybean meal, roasted soybeans, meat and bone meal, fish meal, alfalfa meal, kelp meal (seaweed), brewers' yeast and probiotics (*Lactobacillus acidophilus*). We use no antibiotics, hormones, coccidiostats, synthetic vitamins, germicides and the like.

Stewing hens are from our egg layers who have "served their time." All the comments about these last year were real positive. The meat's taste is very rich, but of course it must be cooked slow and long to be tender. These will be available late in the summer, as production begins to wane.

Ready-to-lay pullets, as usual, will be available around Christmas if you'd like the joy of having a backyard flock for your own eggs. We do not debeak these birds, of course, and raised on range they are good, aggressive foragers.

EGGS

As usual, we will have eggs from our "Eggmobile," the portable henhouse that sanitizes the pasture paddocks following cattle grazing. But we will also have eggs from our "RAKEN" house flock. Remember that we have eggs now, so go ahead and come out for them. Don't wait until May. Spring is the heavy production time. In order to have enough eggs for mid-summer, we always have an oversupply early in the spring. Apparently that's when God intended for us to eat more eggs, so let's join the seasonal roller coaster. Please save us egg cartons. Be assured that these eggs are worth the effort, both for you and for us.

BEEF

A big apology is in order for all of you who ordered beef last year but did not get it. We had orders for twice as much as we had. We warned you, in last year's letter, that we were afraid that would happen. But, the good news is that the shortfall year is behind us and the herd is growing fast. This fall we should have about half again as many as we did last year. We may still be a little short, but we hope it won't be as bad. Next year we should be in good shape. Our beef is grass fattened, raised on perennial polycultures in an intensive controlled grazing program that seeks to mimic the two great herbivore principals: herding (density) and movement (duration). You may notice that we are not offering June beef any more. We believe the fall beef tastes better, having just come off lush fall pasture, and harvesting at that time capitalizes on natural forage and weather cycles. The only beef we may have in the spring is from a cow that loses her calf. In that case, we make ground beef and try to move that to those who want ground beef only. You may note on the order blank if you would prefer to have it in the spring. Remember that the spring, though, is an indefinite deal.

RABBIT

Daniel, at 10 years old, is our resident rabbit expert. Again, our apologies for being so far behind in filling the orders. Two years ago when he had two rabbits and got orders for 150, he thought he'd never catch up. But you know rabbits. He is not feeding them the way conventional wisdom suggests: pellets only. Rather he is feeding them free choice hay, oats and pellets. It's surprising to see what a difference the forage makes in the fat on the rabbit. Of course, rabbit is the only guaranteed no cholesterol meat because its fat stays outside the meat under the hide or inside in the organs. Having gone through the learning curve, he is expanding rapidly, improving the genetics, and produces a fine rabbit, as many of you can attest. Because rabbits produce year-round, he is having trouble matching orders to chicken orders to save folks an extra trip. We have decided to fill orders from you who live far away during the summer when you are coming for chicken. You who live nearby, we hope it's not too much to ask you to come off-season, like fall, winter and early spring, to get your rabbit. It might help out on freezer space too. If all goes well, he should catch up on orders by next spring.

VEGETABLES

Remember that these are available when we have surplus.

FIREWOOD

For heating or romance, wood fires are hard to beat. It's regenerative and doesn't produce the harmful particulates of other fuels. We cut it and pile it here at the house for you to haul at $25 per pickup load. Call us if you want it delivered.

Finally, we're hosting another Field Day this summer, July 11. This is your opportunity to mingle with folks from around the country, see inside our farm and get the scoop on what we're trying to do. A hay-wagon tour with barbecued Polyface chicken and trimmings will add to the day's festivities. Reservations must be made and you Polyface patrons may come at half price ($10 per person, $17.50 per couple, students $5 and children under 12, free).

Please, if you have any questions, call. We now have an answering gizmo. Feel free to bring a picnic and see how we do things. We learn from each other. Thanks again for your faith in us -- we don't take it lightly. Remember the orders are first come, first served, so don't dilly-dally with this thing. Thanks.

THE SALATINS OF POLYFACE INC.

305

Spring 1993

Dear folks,

"I just can't eat that stuff out of the store anymore." We never tire of hearing you say that. We appreciate the seemingly hundreds of times we've heard that statement, and trust we will never violate your trust in producing the world's best poultry, beef and rabbit.

Several of you have called, lamenting the fact that your freezers are well nigh empty and May is still many meals away. We're sorry, but you'll just have to order more birds this year. Cycling our production with nature's seasons must not be compromised. As soon as we produce food against the seasons, disease, costs and environmental degradation escalates. We calve when the deer are fawning, start chicks when wild turkey eggs are hatching, and increase the work load when the days are longer.

We've enjoyed the winter respite, and have been busy spreading our message to other folks around the country. In November, Joel keynoted the Carolina Farm Stewardship Conference in Rock Hill, SC. In early January, he keynoted the Maryland Organic Food and Farming Association annual meeting in Annapolis. The end of January and early February, he spoke at the Texas Eco-Fair in Austin, the Northeastern Grazing Conference in Lancaster, PA and keynoted the West Virginia Direct Marketing Association annual conference in Morgantown. On three of these the family went along, enjoying a mini-vacation.

Many of you know that we have written a 60-page *Pastured Poultry Manual,* and it has sold nearly 1,000 copies in less than 2 years. It has stimulated probably 100 folks around the world to begin raising chickens and eggs using the Polyface model. This winter, we have updated it, increased the information by 50 percent, and added 20 pages of pictures. Lord willing, it will be released as an honest-to-goodness 200 page paperback book sometime in May or June. We've been told that we're starting a revolution in agriculture. Isn't that exciting! You folks deserve much of the shared credit, for standing with us and turning our dreams into reality. Thank you.

We hope this year to put together a brochure about Polyface, an introduction to our methods and philosophy, so that newcomers can quickly understand the game plan and you old-timers don't need to be bored in these letters with unnecessary introductory information every spring. We are taking one step at a time.

RABBIT

Your hunger for Daniel's (the 11-year-old Salatin) alternative rabbit has far outpaced even the ability of rabbits to keep up. He has meticulously recorded all your orders, and knows how far behind he is. For this year, we are not taking orders, but rather focusing all attention on catching up with old orders. His goal is to catch up by year's end, double the breeding stock, and be ready to accommodate new orders next year. Thank you for your patience. Please feel free to take a look at them when you come out.

EGGS AND READY-TO-LAY PULLETS

For the first time it appears we will not be able to keep up with eggs this summer. We had planned to keep at least 100 pullets for layers this year (out of the 350 we started last summer) but the demand for these pullets rose sharply last year and we finally began turning people away when we only had 50 left. As a result, our layer flock will be under 200 this year. We'll squeeze those old gals all we can. Remember there is nothing like having a dozen or so in your backyard to produce your own eggs. We use nonhybrid old American varieties, and encourage you to get some to produce your cackleberries.

CHICKEN

Is it really any different? You bet. We've articulated it this time as an educational tool -- and to help you understand why it's worth a premium price. At the end of this comparison, we'll pick up this letter and close with BEEF.

Quite a list, huh? It is amazing how far off base a production/ processing model built on erroneous beliefs can deviate from what is good and reasonable. If you have any questions about it, please ask. We tried to cram all the information we could on that one sheet of paper.

We always hate to see prices rise without any explanation. You will notice that the chicken price is up a dime a pound, or 8 percent. This is the first price increase in 3 years, and during that time the cost of our chicks has gone up a dime; taxes and insurance have risen dramatically; fuel and utilities (electricity) has increased and our feed bill is higher. We feel badly about having to raise the price, especially with Clinton's pledges to increase taxes, but we must make a profit. Thanks for understanding.

Our last two batches last year were on the small side, and we believe the eggs came from an old flock just going out of production. We are researching alternative chick sources, and if we

308

must change chick suppliers, this will greatly increase our cost of the chicks. For sure, we want those 4-pounders every bit as much as you do. At least you are paying by the pound, so smaller birds are cheaper. You just have to eat more of them. Again, thanks for bearing with us.

BEEF

Again we must apologize for not being able to accommodate all the orders last year. We are expanding as fast as we can. We're coming closer each year to having enough to go around. Everyone seemed pleased with the changes last year at the slaughterhouse, and our meeting you there. We plan to continue that policy. It made Mullins happy, too. Remember that the size differences from beef to beef is quite dramatic since we do not only steers, but heifers also. If you'd like a small quarter, indicate that on the order blank. Do the same if you'd prefer a large one.

You may notice that the price on front quarters has stayed the same, but the other prices have risen slightly. With this huge price spread, the best value for your money is the front quarter. It seems a hopeless cause to spread the price enough to keep fronts and hinds in equilibrium. If we could breed a cow with all hind quarters, we'd really have something -- kind of like a pig that's all ham. But we can't (despite what the biotech monkeys think) or shouldn't, so we're stuck with fronts and hinds. Now please, don't abandon the hinds just because we're making a big to-do about this. We're just trying to explain the price difference and level with you about our needs.

That just about wraps it up for this spring. Fill in the order blank, put it in the enclosed envelope, stick a stamp on, and fire it back to us. Remember it's first come, first served. When we reach our capacity, that's it. We're sure looking forward to seeing all of you again this year. While we do operate a business, we prefer to think of it as a relationship-builder; and while you are customers, we prefer to call you friends. You're welcome anytime. Come for a hike, to work off some flab, to enjoy a picnic or whatever. You are collectively the greatest boss we could want. Fill out the order blank and have a great season.

THE SALATINS OF POLYFACE INC.

WHAT IS THE DIFFERENCE?

POLYFACE CHICKEN	CONVENTIONAL CHICKEN
Unvaccinated	Vaccinated (immuno-suppressant)
Full beak (no cannibalism)	Debeaked (cannibalism a problem)
Probiotics (immuno-stimulant)	Antibiotics (immuno-depressant)
Composting litter in brooder (sanitized through decomposition)	Sterilized litter (sanitized through toxic fumigants and sprays)
Carbon/Nitrogen ratio 30:1	C/N ratio 12:1
Practically no ammonia vapor (smell)	*Hyper-ammonia toxicity
Brooder skylights	*No skylights
Rest at night -- lights off	Artificial lighting 24 hours/day
No medications	Routine medications
No synthetic vitamins	Routine synthetic vitamins
No hormones	Routine hormones
No appetite stimulants	Routine appetite stimulants (arsenic)
Natural trace minerals (kelp)	Manufactured and acidulated trace minerals

WHAT IS THE DIFFERENCE?

POLYFACE CHICKEN

Small groups (300 or fewer)

Low stress (group divisions)

Clean air

Fresh air and sunshine

Plenty of exercise

Fresh daily salad bar

Short transport to processing

Killed by slitting throat (per Biblical directives - see Leviticus)

Carefully hand eviscerated

Processing uses only 2.5 gal. water/bird

CONVENTIONAL CHICKEN

*Huge groups (10,000 or more)

*High stress

*Air hazy with fecal particulate (damages respiratory tract and pulls vitamins out of body, overloading liver)

*Limited air and practically no sunshine

*Limited exercise

*No green material or bugs

*Long transport to processing

*Killed by electric shock (inhibits bleeding after throat is slit)

Mechanically eviscerated (prone to breaking intestines and spilling feces over carcass)

Processing uses 5 gal. water/bird

WHAT IS THE DIFFERENCE?

POLYFACE CHICKEN	CONVENTIONAL CHICKEN
Guts and feathers composted and used for fertilizer	Guts cooked and rendered, then fed back to chickens
Effluent used for irrigation	*Effluent treated as sewage
Customer inspected	*Government inspected
No injections during processing	Routine injections (anything from tenderizers to dyes)
Low percentage rejected livers or carcasses	High percentage liver rejects or carcasses (breast blisters)
Dead birds fed to buzzards or composted	Dead birds incinerated or buried (possible contamination of water)
Sick birds put in hospital pen for second chance -- most get well	Sick birds destroyed
Manure falls directly on growing forage and active soil for efficient nutrient cycling -- converted to plants	Manure fed to cattle or spread inappropriately (ammonia vaporization -- air pollution; nitrate leaching--water pollution)
Fresh air and sunshine sanitize processing area	*Toxic germicides to sanitize processing facility
Cooking loss 9% of carcass weight	Cooking loss 20% of carcass weight

WHAT IS THE DIFFERENCE?

POLYFACE CHICKEN	CONVENTIONAL
Long keepers (freeze more than a year)	Short keepers (freeze only 6 mos. or less)
No drug-resistant diseases	Drug-resistant diseases (R-factor *Salmonella*)
Low saturated fat	High saturated fat
No chlorine baths	Up to 40 chlorine baths (to kill contaminants)
No irradiation	FDA-approved irradiation (label not required)
Environmentally responsible	Environmentally irresponsible (hidden costs)
Promotes family farming	Promotes feudal/serf agriculture
Decentralized food system	Centralized food system
Promotes entrepreneurial spirit	Promotes low wage/time-clock employment
Rural revitalization	Urban expansion
Consumer/producer relationship	Consumer/producer alienation
Rich, delicious taste	Poor, flat taste
Edible	Inedible

*Also applies to nearly all "certified organic" chicken.

ADDENDUM FOR 1996 REPRINTING

Because this is a dynamic, evolving model, we are constantly learning new techniques. In addition, thousands of folks around the country are using their own creativity to refine the model as well. What follows is a list of the latest thinking.

RATION		
	corn	5,000
	wheat middlings	2,000
	soybean meal	1,000
	roasted soybeans	1,000
	meat and bone meal	600
	fish meal	300
	alfalfa meal	400
	kelp meal	165
	brewers' yeast	150
	probiotic	20

Be sure to premix the supplements (from alfalfa meal on down) before adding the other ingredients in order to ensure even distribution.

LEG PROBLEMS   This is a growing problem in the industry. You can topdress feeders with brewers' yeast to increase intake. Another option is to reduce crowding and/or withhold feed for a couple of hours a day from day 11 to 15. Liver fed free choice works exceptionally well.

WATERERS   Instead of using all the plumbing fittings to connect the rubber tube to the plastic bucket on top of the pen, just drill a hole a little smaller than the tube and push the tube in, using a screwdriver to crimp one side to get it started and then a pair of needle-nose pliers to pull it through. Just let it stick through an inch and you're in business -- without any fittings. The pressure is low enough and rubber tube flexible enough to self-seal.

EGGMOBILE   We no longer recommend the *Eggmobile* for acreages under 50. The problem is that chickens have such a strong homing, or nesting, instinct that unless they are kept unfamiliar enough with their surroundings they will find another "home" instead of the *Eggmobile*. You must have enough land area, or at least the land area must be broken up with enough forest runs, ponds and the like, that they will not have access to a stationary "home" like a shop, back porch or barn.

This is incredibly hard to do with a small acreage. Perhaps a couple of field pens scattered about your small acreage and propped up maybe two days per week to allow the birds to free range, would give the benefits without losing them to permanent structures.

To eliminate closing the door each evening, an Amish farmer in Holmes County, Ohio mounted a wind-up alarm clock inside, set to go off about dark. He tied a piece of kite string to the alarm winder and hooked the other end to a stick going through a hasp. When the alarm goes off, it winds up the string, pulls out the stick, and lets the door drop. We are now retrofitting ours -- absolutely brilliant idea! Perhaps electronics whizzes could rig up a radio-controlled unit. Wouldn't it be great to have a little radio-controlled electric motor on the *Eggmobile* wheels so that from your back yard hammock you could move it?

INSPECTION Public Law 90-492 is called the "Producer/Grower Exemption." It is a Federal law that allows a grower to produce or process up to 1,000 head of poultry per year and sell them on the farm and to the HRI trade (hotel, restaurant, institution). The birds cannot be sold wholesale to a retailer -- they must be sold to an end-use entity.

This exempts you from *ALL* government inspection requirements. Roughly 32 states have extended this exemption to more than 1,000, but interpretations differ regarding "inspection." In some cases, a cursory look is all that's required, and in others this is being construed to require mega-facilities. It is a very gray area. Here is what we know:

(1) Bureaucrats do *NOT* know what the law is. They interpret it to suit their political fancy. All kinds of loopholes exist; you and your spouse each raise 1,000, add some customers to raise 1,000. Use a pre-buy arrangement so you do not own the birds. Sell by the head instead of by the pound. Drive the tyrants crazy.

(2) The issue is political, *NOT* a matter of public health or quality. This battle will be fought in the political arena. It is a freedom of choice issue. If you preclude entry-level entrepreneurs from entering the marketplace, you inherently deny creative alternatives to fecal soup. Enlist help from your elected officials, media and patrons.

(3) Inspectors are spread very thin. In many states fewer than half a dozen are responsible, and this means as more and more of us do this, they can't keep up with us.

(4) Don't ask, don't tell. The worst thing you can do is ask if you can do this. Of course the bureaucrats will cover their trail by telling you "no." It's easier to ask forgiveness than permission.

(5) Major change, real societal headway, always comes with a price. Securing a beachhead is
costly to the first wave of troops, but look what
it achieves for the second group. If we do not
passionately establish a beachhead on this issue,
our children will have no option but genetically
engineered, irradiated, chlorinated, adulterated
fecal soup. Who will join me in this great ministry for clean food?

ANECDOTAL FINDINGS   Two years ago we submitted
our skinless chicken breasts to the food sciences
laboratory at Virginia Tech for fat analysis compared to conventional supermarket birds: Ours were
70 percent lower in fat. Of the fat present,
saturated and monounsaturated were both lower while
the good fat, polyunsaturated, was significantly
higher. Other comparisons from around the country
are yielding the same ratios. The "organic-ness"
of the ration has nothing to do with these findings: the fresh daily pasture intake is the critical factor.

Also two years ago, two microbiology students
from James Madison University took skin samples to
measure bacterial contamination. Measured in colony-
forming units per milliliter, the conventional supermarket birds (which had been chlorinated multiple times) averaged 3,600 CFU/ml and ours were
133, a *25-fold difference!* Yes, there is a difference. Let's give people some real food for a
change.

Addendum for
Pastured Poultry Profits,
1999 Reprinting

Each year new difficulties arise that sometimes make us wonder how long we can continue to raise industrial birds on pasture. As the industry selects for certain genetic traits for a production model going 180 degrees different than ours, we must be more and more creative in order to get satisfactory performance.

In the last couple of years we have actually wondered how long we can continue to use regular hatchery birds and the "normal" feedstuffs. The quality of both feed and birds has deteriorated dramatically in the nearly 15 years we've been raising these broilers. The industry selects for birds that can gain faster on higher calorie feeds. To compensate for the physiological strain this puts on bones, tendons, organs and nerves, the industry develops ever stronger medications, hormones, arsenicals and vaccines.

Each year the birds need higher octane and more engine fine tuning in order to perform, but our paradigm calls for leaded, stodgy fuel and no mechanics to tweak the engine every few hours. As we and the industry move farther and farther apart, we find it nearly impossible to get good performance from the same ration we've always used.

In addition, when we started people weren't feeding dead cows to cows and you could get rendered animal proteins from fairly pure sources — all cow, for example. But over the years, pure-species animal protein has become virtually impossible to get. Every one on the market contains some poultry by-products, which pushes us into feeding industrial dead chickens back to our chickens.

Not only is that unnatural and philosophically reprehensible, it is the slippery slope that has led to mad cow disease in Britain and who knows how many animal diseases.

We finally realized last year that we could no longer feed just "natural" things to these chickens and have them perform; we needed to upgrade the octane with some high-tech biologicals. Obviously, we didn't want to use medications and synthetics, but there are some high-powered biological extracts, some genuine hyped-up non-synthetic snake oil. We simply could not get enough vitamins and minerals into these birds from natural feedstuffs to get decent performance because these birds were completely different than the ones we had a decade ago.

As I travel around the country, I ask a lot of questions to see what I can learn from folks who are creatively refining this pastured poultry model. In addition, most conferences have trade shows with every kind of nutritional product and promise you can imagine. The problem is that all these product sales reps show charts that compare their product to nothing.

It's always "ours" compared to "nothing." But you and I don't make decisions that way. We need to decide, among several options, which is best.

But these products don't compare themselves to each other; the control group is always "nothing." So which snake oil should I buy?

We've tried numerous things over the years. We tried Nutri-Carb and nearly killed half a batch of chickens. We tried hydrogen peroxide, with no results. Over the years, when someone promised a great solution, we'd try it. That's how we came up with the Conklin Fastrack probiotic — others were just glorified minerals with enough bugs thrown in to be called a probiotic. The Fastrack really gave us results in reduced late-growth heart attacks.

We resolved, therefore, to run trials this year comparing these competing snake oils with each other, dividing our flocks and running these products head to head to see which would give us the best performance. We measure performance first in terms of mortality and secondly weight gain. Since we don't mask sickness with medications or vaccines, health is real and not a charade.

The first batch of 1,300 this spring we divided into four groups of 325. One received our regular old ration. Group two received the regular ration plus Immuno-Boost, a water supplement. The third group received the regular old ration plus Willard's Water, a catalyst altered water supplement. The fourth group received no water supplements but rather a ration and feed supplement (Nutri-Balancer) containing NO animal protein developed by Fertrell, an organic soil amendment company that has been in business for nearly 30 years.

Very shortly we noticed that all three supplemented batches spun circles around the control. This was enough to prove that our hunch about needing to look at some snake oils was well founded.

All three of these groups performed extremely well. Imagine how elated we were to find the Fertrell Nutri-Balancer group doing so well without any animal protein except some low-heat fish meal (Sea-Lac).

As we went to the field, things rocked along but the Immuno-Boost clogged up the waterers terribly and we abandoned that product. With the Fertrell birds doing so well, we finally switched everything to that and did not run the trial completely to the end since we are in the profit business, not the pure research business.

Incredibly impressed with what we were seeing from the Willard's Water, we broke the next batch into a group of Fertrell only and Fertrell plus Willard's. Of course, we joked about having 10-pound birds in 2 weeks and all the stuff that goes with experimenting with snake oils.

Interestingly, adding the Willard's Water to the Fertrell did not make an iota of difference. We determined that we could not make the birds "weller." Once they are well, they are well and that's that. Now we've abandoned the Willard's Water and intend to try some other snake oils to see if we get any benefits.

We are pleased enough with the Fertrell supplement that we have put all layers and broilers on it as our new "control" and will begin running trials from this new benchmark. We also want to give some competitors a shot at the ration, specifically Leland Taylor and his Clodbuster stuff, as well as Dynamin and others — do you have ideas?

I really believe that this type of research is sorely needed, but it cannot be funded by companies or institutions or it can be biased. All I want is what performs, plain and simple. I have purchased, at full retail price, every ounce of every product I've used. Any trial that uses "free" product should be questioned for integrity. I am not trying to hurt or help anybody except to share our experiences with practitioners who are smart enough to make their own decisions.

For sure, we do not ever intend to go back to animal proteins. The theory is that the minerals cause the body to metabolize the proteins in the grains. As we've reduced the minerals in our soils through chemical fertilizers, we've reduced the vitamin-mineral content of our grains and hence the body's ability to extract the nutrients that are there. By properly supplementing with minerals, it makes everything kick in like it should. Pretty fascinating. To think that the whole protein issue is really a mineral weakness really makes you shake your head.

It's also fascinating that the other two water supplements, on our old ration containing the meat and bone meal, had incredibly dramatic results, almost miraculous. Certainly if a person were using conventional rations, I have no doubt that using either of these products would solve a multitude of problems and pay for themselves many times over. But finding the crutch that fits is not as good as healing the broken leg. I think the ideal is to get completely away from the industry as much as possible, and for sure a good place to start is with meat and bone meal.

Two weeks ago we took our second batch to the field, 1,053. We could not find one cripple or a single gimpy, weak bird. Since then, we've had several frosts and have not lost a single bird. They are clean and white, with the pinkest skin

I've ever seen — wonderful color. The chicks have clean rear ends — they look almost shampooed and air-brushed instead of manure-smeared as is common. Of all the things we tried, I certainly did not expect an animal protein-less feed to work, but it has made a believer out of me. We no longer need a hospital pen. The company says this product is organic certifiable.

Our current layer ration is as follows (in pounds):

corn	5,000
roasted soybeans	3,100
oats	1,100
feed grade limestone	500
Nutri-Balancer	300
Thorvin kelp	55

Crimping the oats and cracking half the corn adds texture and makes the feed less powdery. This layer ration has definitely decreased old-bird mortality, although production does not seem to be affected.

To really get chicks off to a good start, try feeding about one dozen boiled eggs per 300 chicks per day for the first week. Just mash them up, with shells on, and put in on newspaper or sprinkle on top of feeder. This is also a way to make sure turkey poults jump up and take off. This is a great way to use pullet eggs or cracks from your egg laying operation.

1999 Addendum

What About Eggs?

When the original *Pastured Poultry Profits* came out we were only producing eggs in the "Eggmobile" and the "Raken House." The "Eggmobile" is a portable henhouse that we move a couple of days behind the cattle to sanitize the paddocks. The chickens free range from the trailer, eating grass, bugs and especially fly maggots in cowpies, spreading out the pasture droppings in the process.

The "Raken House" is a combination chicken house/rabbitry, which was originally built to house our son Daniel's rabbit operation. A loose-housed laying hen facility, it has two tiers of production: laying hens on the floor and rabbits in cages at eye level. The chickens scratch through the rabbit droppings and aerate the bedding, producing wonderful compost and a delightful woodsy odor.

About three years ago, we added a commercial-sized pasture egg operation to the farm when we started the apprenticeship program. After all, we needed something for the apprentices to do! Actually, we needed an enterprise that would cash-flow the added expense of teaching, housing and feeding a couple of young men.

We built a fleet — 29 to be exact — of field pens for the pastured egg that are almost identical to the broiler pens. We put an 18-inch-wide hinged door on the top of the pen, all the way across the 10-foot dimension, that would provide access to 10 nestboxes tucked in the top corner of the capped end. Although the door creates a crack in the roof, any rainwater falls in front of the 12-inch deep nestboxes and the layers do not mind.

The nest boxes have a 6-inch front board to keep the birds from scratching out nest material and a 6-inch partition board on 12-inch centers to individualize the nests. Without partitions, the birds all lay in one spot and break the eggs. A perch board in front drops down in the morning and comes up in the evening to prevent the birds from roosting in the boxes and soiling them at night.

We had to build a dolly with shorter prongs to move these pens because the nest bottoms are only 12 inches off the ground. The bottoms are simply pieces of corrugated aluminum or sheet metal — plywood is too heavy.

It takes as long to build in the boxes as it does to frame the entire pen. The boxes add significantly to the weight of the pen. Anything to lighten them — using plastic boxes or whatever — can help.

We put 50 birds in a pen and shoot for at least 36 eggs per day. That is fairly easy during the first year of production, but is impossible the second year. We like to keep the birds two years because one of biggest hurdles of the egg enterprise is dealing with the spent hens. We are actively cultivating a market for this new product

— stewing hens hardly exist now because the industry so debilitates the birds that in 10 months
producers must pay someone to come and haul the
spent hens away.   Ninety percent of second-year
eggs are jumbos.

Anyone over 50 remembers "chicken and dumplings" and eagerly tries one of these birds.   But
younger folks have never heard of this critter.   As
a result, we're in the "giving samples" stage of
our marketing on this one, but the results have
been extremely good.   Our goal is for outgoing
flocks to pay for incoming ones.   Stewing hens
average a little more than 3 pounds carcass weight;
replacement costs a little under $3 per bird (90
cents for the chick, plus $1.80 in feed, plus la-
bor).   By charging $1 per pound, we can accomplish
this goal, which is critical to an economical egg
enterprise.   The average laying operation uses the
first 3 months' production to pay the capitalization costs of the replacement birds.

But if you can pay for the new with the old,
you can begin making a profit early in the production cycle.   Here are our most successful uses for
stewing hens:

♠   **Precooked:** Put as many birds as possible
in a large roaster pan (do not add water), place
in the oven at 325 degrees, and cook for 4-6
hours (until the meat pulls away from the bones
and/or falls away from the bones).   Pick off the
meat, dice it into freezer containers, and freeze
for precooked chicken ready for quickie casseroles and such.

♠   **White tablecloth restaurants** live and die
on the quality of their "stock."   These stewing
hens are made to die for among good chefs, and we
have found great receptiveness.   The broth is

unparalleled. The meat can be used in chicken
salads.

♠   **Use as congratulatory freebies.** All summer, we've been giving away a stewing hen to
anyone who purchases more than $20 worth of goods
at our Staunton/Augusta Farmers' Market booth.
Not only does this stimulate sales, but it also
acquaints folks with this brand new product.

Of course, dressing these birds is much more
difficult than broilers. The tissue is tough and
the insides are full of strange things. Other
outlets we've heard of but have not yet been able
to tap:

♠   **Fox hunts.** Sell the local fox hunting
group your stewers for $2-$5 per bird (whatever
you can get) and put them in some little chicken
huts scattered around the hunt area, just to feed
the foxes. It brings in foxes and that makes for
more exciting hunts. The chickens can just free
range from these range shelters.

♠   **Gourmet Chinese restaurants.** We know one
man who is getting $2.25 per pound for his stew-
ing hens with the head and feet still on! Remember, Orientals appreciate colored poultry.

♠   **Zoos.** We sold a few old hens to a snake
show at the county fair this summer. He let us
watch a 200-pound Burmese python kill and eat one
— quite dramatic. It struck us that zoos must
spend huge amounts of money on food for carni-
vores, and this would be exciting because they
can release live animals — chickens!

Marketing the pastured eggs has been one of
the most enjoyable things we've ever done. The

quality is unbelievable. Just buy a dozen eggs out of the supermarket and crack a pastured egg next to it — you'll think twice about ever eating a store-bought egg again.

Now a few more nuts-and-bolts about economics. We've pushed a pencil pretty hard on these three models: loose-housed, portable house/free range, and portable field pen confinement. Each has assets and liabilities.

The loose-housed asset is labor. Feeders and waterers can be virtually automatic because they stay in one place. Nothing to move. Of course, the downside is that pasture is unavailable and a yard, regardless of its size, will eventually turn to dirt unless it is rotated frequently and even then it must be downright expansive to keep from becoming unvegetated. A yard and fence that expansive is horrifically expensive.

We have tried greenchop, like putting in lawn clippings and such. The problem is that, unlike a cow, a bird's metabolism is extremely high and storage is low. While an herbivore can eat and hold nearly 20 percent of its body weight to be digested slowly, a bird can only hold less than 5 percent. This means that a bird must eat smaller amounts more frequently. What happens is that the chickens dive into greenchop but in a matter of minutes eat all they can hold and go off to do other things. By the time they come back to eat some more, the forage is wilted and they don't want it.

We have tried some fairly expensive dehydrated alfalfa cubes from Canada, reconstituting them in water, to see if the birds will eat them better, but that has not worked. One thing seems to hold some promise and I hope to pursue it:

sprouts. The problem, of course, is that this may not save any labor over the other two models. The house also requires large amounts of bedding carbon (litter) and a relatively expensive structure.

The death knell of the house is capped bedding. This occurs when manure load exceeds the birds' capacity to scratch it into the bedding. Our experience indicates this occurs at the 3 sq. feet per bird level. At 5 sq. feet per bird, capping is not a problem at all. Offering that much floor space is expensive; hence, the multispecies idea helps make it economical.

The portable house/free range model offers the highest quality egg from a nutritional, taste and overall performance level, but moving the trailer requires a tractor or pickup. You really cannot start a tractor, drive out, hitch up, move, unhitch and drive home in less than 30 minutes. In that amount of time, however, you can feed and water 1,000 loose-housed birds or move and service 500 birds in field pens.

The "Eggmobile" offers the cheapest egg since the birds pick up 70 percent of their ration off the pasture, but loses the greatest amount of birds. Some wander off and never come back; some get picked off by predators; some stay around the house and poop on Mom's carport, thereby losing their status in life. The point is that the model that offers the greatest freedom also loses the greatest amount of control.

In our 12 ft. x 20 ft. "Eggmobile" this year we put 270 layers — by far the most we've ever tried — and they did great. Next year we plan to go to 400 with a greatly modified design. Stay tuned. Since the moving and servicing is a fixed overhead, the more we can move per hookup, the most economical it becomes. Increasing the birds is paramount to making this model commercially viable.

One of our previous apprentices, Tai Avanzato, solved the "teach them to go in at night" problem by wrapping poultry netting around the bottom of the trailer the very first night. Voila! The birds went right in — major breakthrough. Right now we intend to put all second-year birds in the "Eggmobile." That way, their low laying percentage won't matter as much because they eat so little purchased grain, and the birds we do lose will be old ones.

Now to the field pens. They can be moved efficiently by hand — although they are much heavier than the broiler pens. They provide wonderful control and protection. They offer a hybrid of the loose-housed but with fresh pasture. The downside is that they require lots of labor and net only $3 per pen per day compared to $7 per pen per day with broilers. And if the birds do not lay more than 50 percent, they don't pay the feed bill.

The loose-housed birds give us the best whites; the pastured birds give the best yolks. The "Eggmobile" birds always have dark, rich yolks but the field pens fluctuate with the season. In the spring the birds eat forage readily but in the hot summer they do not consume as much of the tougher grass. The field pens require no bedding like the loose-housed birds and that saves labor and money. However, the field pens are half a mile from the house and necessitated purchasing a four-wheeler in order to bring in the eggs. Now that we have one, of course, we use it all the time for other things.

Both the field pens and "Eggmobile" need to be shut down in our winters — not just for the chickens' sake, but for ours. Carrying water buckets around to 20 pens in 3 feet of snow is not my idea of a picnic. We built two 20 ft. X 120 ft. hoophouses for winter housing and they have worked great. In fact, Daniel moves his rabbits in as well and one house has an earthworm bed going down one side, under the rabbit cages. Heavy wire cloches on top keep the chickens out but encourage the chickens to roost and poop during the day, further feeding the earthworms and reducing the manure load on the bedding.

By wetting the bedding along the earthworms, we stimulate worms to crawl out into the bedding, where the chickens scratch and eat at will. In the spring, we put vegetables in the bedding and get multiple use out of the structures.

I've been reluctant to say too much about the egg enterprise because we're still very much on the front end of the learning curve. I think it will be another couple of years before we feel comfortable with the model or combinations of models that prove most efficient.

In the summer of 1998 we developed a hybrid egg production model that appears to be the next quantum leap for this enterprise. Premier, the electric fencing company (phone number 800-282-6631) has a new poultry netting material similar to flexinet, which is used primarily for intensive sheep control. This electrified netting has built-in fiberglass posts, is 42 inches high, and comes in a 150-foot roll.

We built a crude — can you imagine me doing anything else? — A-frame out of two 20-foot locust poles and salvage roofing measuring 20-ft. x 15-ft. It's roughly 6 ft. high at the apex and we fastened regular metal nest boxes along the side. That way you can walk down the middle, where the roof is high enough, and gather eggs along the side, inside. A little feed sled chained to the back carries about 300 pounds of feet in three metal feeders underneath a hinged roof to protect the feed from rain.

The range shelter can be any type of structure that is portable and rainproof. Skids work well because the structure will not roll on a hill, like a wheeled structure would. Because chickens are not tall like cows or horses, structures need not be tall or vertically-sided.

This year we are planning to build 20-foot x 20-foot hoophouses on skids as the next improvement over the A-frame. We can cover the hoops with transparent woven poly and then cover that with silvered reflective poly (all available from Northern Greenhouse Sales at 204-327-5540) for summer. In the winter we can simply take off the reflective poly and have a solar hoophouse. Since we plan to have four of these, we can push them together for a longer but segmented hoophouse for winter housing. This will give us year-round use.

We clipped one wing on each bird to eliminate flying and put them in this netting circle, electrifying the material with a Speedrite 12-volt charger powered by a regular deep cycle marine battery. The Premier 18-inch ground rod was adequate to get a 6,000-volt charge. Three rolls of the netting encircle almost a quarter acre, which is big enough for about 1,000 birds for three days and still give them three times as much area as they have in the pens.

End View

regular rafters, purlins and
steel roofing for sides

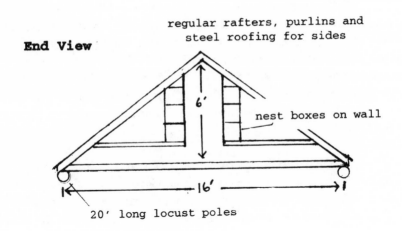

6'

nest boxes on wall

16'

20' long locust poles

Bird's Eye View

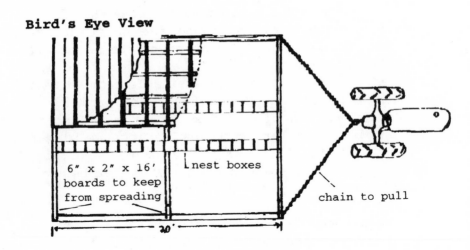

6" x 2" x 16'
boards to keep
from spreading

nest boxes

chain to pull

20'

A-Frame Layer Range House

drawings by Charles Dunaway

334

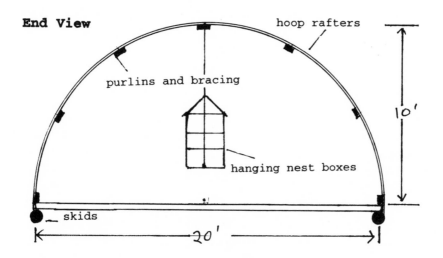

End View

hoop rafters

purlins and bracing

hanging nest boxes

10'

skids

20'

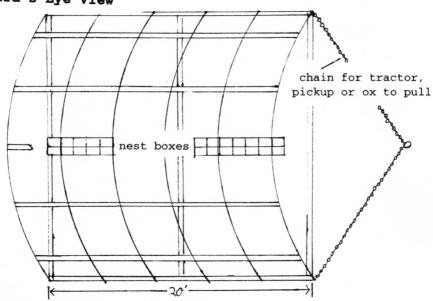

Bird's Eye View

chain for tractor,
pickup or ox to pull

nest boxes

20'

New Prototype Range Hoophouse for Layers
drawings by Charles Dunaway

335

We purchased enough netting to have two complete circles, which allows us to open one and move the A-frame and attached feed sled into the next circle without letting the birds get out of the enclosures. It takes about one hour to move the whole shebang.

The bottom line on this new development is that it saves us two-thirds in labor by not having to move all the pens, it makes better and cleaner eggs due to the additional pasture allotted to the birds, and it reduces our feed costs by about 20 percent since the birds are eating more off the pasture. All these changes are positive and we are now planning to double our egg production since we can now put more time in gathering and packing eggs instead of daily pen moving. The old pens we're just renovating back to broiler and turkey pens. This model only works for layers, not broilers.

This new model offers a couple of advantages we hadn't thought of before. First, the netting protects the whole affair from cattle. Whenever we ran other stock in the field, we would have to put a temporary fence around the pens and hope a deer didn't go through it while the cows were adjacent to the pens. But with this system, the electrified netting makes everything inherently cow-proof, without any additional fencing requirements. This means that this pastured egg cell could be run on any livestock farm worry-free, opening up many new opportunities for wanna-be farmers.

Secondly, the fencing keeps predators out, eliminating the evening close-up requirement for the eggmobile. It gives us the range advantages of the eggmobile but the security and control of the pens. Whenever we bring the eggmobile up around the house, the birds get in the flower beds and other places we don't want them. But this system gives us a compromise, kind of the best of both worlds.

One additional thing we did do in 1998 was build a second eggmobile with a rear hitch so we could hook the two together. We've pushed our numbers up to 400-plus in these 12-ft. x 20-ft. eggmobiles by adding another door and ramp on the opposite end. This way birds have two entrances. Now we can run 800 birds in the eggmobile train, which makes the daily move much more efficient. Now for the half-hour invested in driving out with the tractor and hooking up we move 800 birds instead of just 200 or 300.

Many people ask why we don't just build more eggmobiles instead of going to this netting cell arrangement. The reason is that the eggmobiles are a land extensive system, and with only 100 acres of pasture, we can't keep them far enough apart to insure flock separation without having them in tall grass or areas we may not want them. We don't want to run them through tall, grazable grass because they mash it down and waste it. Since the birds range out 200 yards, a minimum 200-yard separation from the outside grazing area is necessary in order to keep the flocks from mixing. That means it would take a field 600 yards long just to have the two sets in the same field, and we do not have fields that big. Additionally, the eggmobiles do not give us the control we'd like to have — both for protecting the birds and keeping them out of certain areas.

The netting gives the control along with the land intensity, plus additional pasturing and reduced labor. One nice thing about just moving them every second or third day is that the maintenance is more flexible. One of the keys to efficient farming is to minimize the daily chore-time requirement. If we move them the second day, or get tied up and let them go an extra day, it's not a big deal like it is with the pens. Furthermore, we move the pasture cell in the afternoon rather than in the morning, which takes the pressure off chore-time. Less shaded area in the egg cell arrangement means more efficient grazing of the available square footage.

The broilers will not work in this arrangement because they are too small, too stupid and too vulnerable to everything. The mature layers, however, have enough sense to go into the shelter when it rains or when a hawk comes overhead. We are looking forward to more improvements in the pastured egg model, but this latest development has definitely been a huge step forward.

Happy egging.

1999 Addendum

Hoophouses

Several years ago when we added nearly 2,000 laying hens to our pastured poultry enterprise we faced the challenge of winter housing. The six-month on, six-month off seasonal broiler production model suddenly became woefully deficient in a year-round egg production model.

I had never set foot in what I considered a really acceptable stationary confined poultry house. But leaving the birds out in pens all winter was equally unacceptable. Although we knew the birds would survive, the logistics of wading through 3-foot snow drifts with egg baskets and water buckets did not seem like an exhilarating way to farm.

Hoophouses came to our rescue. With each passing year, we are more confident that these are if not the best, at least the best solution we know about.

Chickens only lay eggs after all other energy needs are met. If they use all their caloric intake to maintain body temperature, they do not have enough energy to produce eggs. If a house is tight enough to accomplish this with the chicken's body heat, it lacks ventilation. Supplemental heat sources like wood, petroleum or electric heaters are too expensive.

Hoophouses use solar energy, and even though they cool down at night, the chickens snuggle together to stay warm and the cold does not affect them too much. You and I can handle a cold bedroom as long as we can snuggle down under plenty of blankets. But in the day, when we are moving around working, eating and relaxing, if we can't get warm we are soon miserable. Surrounding, or ambient temperature, is more critical during our production time.

Hoophouse kits are available from a number of suppliers. We purchased one and the pieces came in on a commercial truckline. We opted for the cheapest per-square-foot model that was still high enough to back a tractor or truck under. This model is 20 ft. wide and nearly 12 ft. tall at the apex. The hoops slip into 3 ft. columns anchored in the ground.

We knew the house would need deep bedding for proper manure management. That required hefty sideboards because otherwise the bedding would just push out underneath the plastic. We opted to split locust posts and nail them on upright stakes so we could have 18 inches of bedding in the facility. Then we went along just outside this containment wall and pounded in the 3 ft. pipes, called columns.

In the instructions, these columns are supposed to be in concrete, but we just pounded them into the ground with a sledgehammer. In order not to peen over the top, we made a sock to pound on out of bigger pipe and a heavy piece of metal. A friend who duplicated our efforts used a pipe duct-taped to a piece of wood. A minute of dressing with a round file readies the column to receive the hoop pieces.

We use webbed 9-mil. poly for the cover because it is tough. Although it is more expensive than regular hoophouse plastic, it can last up to 15 years. It is the same material used along the open edge of confinement poultry houses. It rolls up or down depending on ventilation requirements. Using self-drilling screws, we attach a rot-resistant wooden firring strip at ground level. Then we pull the plastic to another piece, roll it up tight, and drywall screw the wrapped piece to the stationary piece. It's poor-boy, but works quite well.

Because the chickens can easily scratch the plastic and rip holes in it, we fasten a firring strip about 6 ft. off the floor along the inside of the hoops. We attach 6 ft. x 1 in. poultry netting to the top of the bedding containment wall and this wooden strip, keeping the chickens from being able to access the plastic.

Because chickens move and plants do not, these hoophouses need both ventilation and a physical barrier. As a result, we put in four locust poles at the ends of the hoophouse: two are 12 ft. apart for the door and two are 6 ft. beyond those in either direction to hold a sliding door rail. The sliding door allows us to regulate the ventilation from fully open to cracked, one end of the house or both ends.

Swinging poultry netting doors, mounted with conventional strap hinges and pins on the same inside locust poles, keep the chickens in when the sliding doors are open. To keep the bedding in we just place a couple of retainer boards across the 12 ft. opening. When we clean out with the front end loader, we just pull off those retainer boards and drive in. Ditto for putting bedding in.

Buying everything brand new except the locust poles, these structures cost us $1 per square foot to build and cover. We orient them so that the prevailing winds hit the end, thus stimulating ventilation on warmer days. We use regular hanging waterers, nest boxes and trough feeders. Even though the waterers freeze on exceptionally cold nights, as soon as the sun comes up they thaw and begin to work.

If we are in a period of single digit temperatures or below 0 degrees, we keep a couple of those indestructible black rubber pans to give the birds water until the sun warms things up enough. A frost-free hydrant in each house, connected to the farm well, gives us plenty of water.

Chickens do not like things damp, so I suggest making sure that any runoff from the hoophouses can make a quick exit away from the house and not go into it. We have two hoophouses 20 ft. x 120 ft. The houses are about 3 ft. apart. We've had snow piled up level to the top of the houses in this crack, and that is a lot of water when it melts.

In houses birds drop just as much manure as they do when out on pasture. Ideally the birds will stir their carbonaceous bedding faster than they cover it with manure: this stirring encourages slow decomposition and ties down ammonia. We have found that at 3 sq. ft. per bird the manure load is heavier than the birds can incorporate, causing a capping of manure on top of the bedding. Once the bedding is capped, the chickens do not scratch through it.

At 5 sq. ft. per bird the manure load is light enough that the birds have no problem keeping the bedding clean with stirring. At 4 sq. ft. some areas may cap and others will not. Capping can be eliminated with something as simple as a mattock or garden hoe or as elaborate as a garden tiller (do it at night when the birds are sleeping).

Another technique we've used is pigs under 100 pounds in 6 ft. x 10 ft. x 3 ft. portable pens that we just walk through the house. The pigs aggressively root up the bedding too tough for the chickens to till. Use only small pigs; large pigs enjoy chicken dinners too much. Two pigs per pen seem to work well to stir up the bedding without re-compacting it. The chickens enjoy roosting on the pen and eating earthworms the pigs dig up.

The raw material for the bedding can be any-thing carbonaceous, but I would stay away from anything with long fibers like old hay or straw. The best is sawdust, although wood chips work well as do shredded leaves. Anything fluffy and friable works fine.

To stimulate the chickens to stir the bed-ding, we fling whole wheat on the floor each morn-ing. The chickens enjoy the treat (about 20 pounds per 300 birds) and invariably they miss some that ends up in the bedding. Some of these kernels actually sprout in the bedding, offering sprouted grain to the birds when the pigs expose it.

Clean bedding is the key to clean eggs. Keeping the nestboxes well away from waterers gives the birds plenty of time to clean off their feet before entering nests. Our birds seem perfectly content to sleep on the floor since that's the way they sleep in the field. Certainly roosts could be provided, although that would tend to concentrate manure. We keep sawdust or bags of peanut hulls close by to spot treat manure concentration areas.

We find that it takes about 50 days for the carotene from the pasture to wear off in the birds' bodies. This makes yolk color suffer the last 50 days of the housing period. We fortify the ration with about 2 percent alfalfa meal to push more greens into the birds. Of course, production is never as high during the winter as it is at other times during the year so we need not worry about sales pitches to new prospects during this lower quality time.

We did try some marigold petal dust from Mexico one winter. Even though it did darken the yolks, the color appeared brownish rather than the more natural orange. Our customers recognize that the color will change and simply accept it. One of our gourmet chefs says that when he was in chef school in Switzerland, they had seasonal specialties to capitalize on the color and handling quality differences of the eggs as they changed throughout the year. He tells me: "They talked about April eggs and May eggs and June eggs, and adjusted the recipes accordingly."

Certainly large-scale sprouting is something we need to be looking at. But since we haven't done it I won't act like my current ideas will work. We'll have to wait and see.

Fortunately in the spring it only takes about seven days for the pasture yolk color to come back up in the eggs once the birds hit tender spring grass. At the first sight of green, the birds can go out. The early grass seems almost like a tonic to them, and is by far the most palatable forage they encounter for the whole season.

After the birds go out, we plant summer vegetables like tomatoes, squash and corn in the hoophouses. This gives additional cash from the hoophouses and allows us to get a jump on outside-planted varieties, insuring a premium price.

We never clean the houses down to the dirt floor, but skim off only what is necessary to keep the bedding below the retaining wall. A midpoint division in each house gives us four sections. Not only does this allow us to keep flocks separate, but it also reduces the number of birds in any one group. This reduces stress and increases production.

The hoophouses keep the birds warm and productive. Even when the temperature is below 0 we have no problem maintaining 70 degrees or better. The plastic allows maximum light, further stimulating production. The direct sunlight keeps the bedding dry and reduces pathogens. Plants double its usefulness and the entire structure, at $1 per square foot, is cheaper than any other weather-proof building, even if all you want to do is store machinery.

Hoophouses properly modified for chickens offer a cost efficient, enjoyable winter housing model. A retired neighbor came over one day after we got birds in the houses and stood one Sunday afternoon with his wife for at least an hour, just watching the birds and enjoying the sunlit ambiance of the facility. All he could do was shake his head and recount the smelly, dusty, dank houses of his youth. He was visibly entranced by the notion of chickens in a hoophouse, and made me appreciate the discovery even more.

The hoophouse model is so appealingly comfortable on cold winter days I think I'll move my office out there. I wonder what typing method chickens use on a computer keyboard: hunt and peck?

1999 Addendum

Turkeys

In the last two years we added turkeys to our poultry portfolio and from what we've already seen, we could have another enterprise fully as profitable as the pastured broiler. In 1997, we produced and processed 350.

First, let me give credit to one of our apprentices, Tai Lopez, for pioneering the first year and developing this addition. We encourage our apprentices to develop new enterprises under the shelter of our umbrella. This builds their confidence for a repeat performance on their own and it stimulates us to diversify by simply maintaining an ongoing endeavor instead of doing all the research and development.

Turkeys are both similar to and different than chickens:

BROODING. Chickens need 90 degrees for the first couple of days and by three weeks can handle freezing if they are gradually hardened off. Turkeys need 95 degrees for the first few days and cannot take freezing until at least 5 weeks old. They can take 50 degree nights as early as 4 weeks, but not at 3 weeks. Hold them inside an extra week or two compared to chickens.

RATION. We use one ration at about 19-20 percent protein for our broilers, throughout their 8 weeks of life. Turkeys, on the other hand, need a 28 percent protein for their first 5-6 weeks. In the wild, poults only thrive on open ground where they can eat plenty of bugs. From weeks 6-10 they can do fine on the broiler ration. After week 10, you can drop down to a 15 percent ration by upping the corn percentage. One way to jump up the protein early on is to feed eggs for the first few weeks. We just hardboil them and mash them onto the top of the feed, shells and all. The poults quickly learn to devour this supplement and it acts as a tonic. Feed no more than 25 percent of the ration.

PERSONALITY. Turkeys are much more people-friendly. Whereas chickens tend to flit away as you come close, turkeys actually come to you. In fact, folks who have tried complete free range on turkeys complain that it doesn't work because the turkeys keep following them back to the house. They just love people and that makes them enjoyable for children when the poults are small.

FORAGING. Turkeys will eat up to 50 percent of their ration in forage, as opposed to chickens where 20 percent is excellent. This means that the more ground the turkeys can cover, especially if it is succulent forage, the less feed they will eat. Their long legs allow them naturally to cover more ground. They simply eat grass more aggressively. Even 10-day-old poults will eat 10-inch blades of grass. Poults readily consume lawn clippings or handfuls of grass. To stimulate foraging and reduce feed costs, we drop the bird numbers to only 12-20 per pen and move them twice a day the last month. This also means that the difference between pastured chickens and their supermarket counterparts is not as noticeable as the difference between pastured turkeys and their supermarket counterparts. The differences are simply incredible.

GRIT. Part of the trade-off for being aggressive foragers is the requirement for more grit. Turkeys consume prodigious quantities of rocks. You can buy this poultry grit from a feed store, but we just scoop gravelly material from sandbars along the creek. As the birds get bigger, they will eat rocks the size of marbles. We just go along once per week and drop a half-gallon or so in a pile on the ground before we move the pen. This procedure keeps the turkeys from being frightened like they are if you drop something strange into the pen. They find all the little rocks and eat them readily.

EMOTIONS. Turkeys enjoy a routine more than chickens. Although they have a bad reputation for stupidity, I think turkeys are much smarter than chickens. When I go out to move the pens they see me a hundred yards off and begin to pace the front of the pen, talking wildly about what is about to happen. Anything out of the ordinary sends them into a panic. Even putting some whole corn in the feeder will make them flutter to the far side of the pen and refuse to come to the feeder for half a day. Their ability to respond positively to routine is the same trait that makes them panic at any routine change.

LIGHT. Turkeys do not like shade or shadows. When we first began raising turkeys we noticed that they would not graze or rest in the shaded cap end of the pen. When we moved the pen, we could not even find any manure there. Effectively, they only used half of the pen. So this past summer we made half a dozen pens without any roof or sides—just poultry netting. Only the door over the feeder was solid. Everything else was wide open. The turkeys immediately spread out, grazed evenly, and performed superbly. This distributed manure evenly as well. Obviously, these pens were extremely cheap to build and lightweight. The Thanksgiving birds went into our regular broiler pens because they needed the shelter, but by that time the sunlight was more slanted so rays penetrated farther into the pen.

HEAT. I know you might be thinking: "If you took off all the shade, what in the world did the birds do when it got hot?" Good news, folks. Turkeys in small groups seem to handle any heat just fine. Oh yes, they will pant just like a dog or any animal, but in blazing 90 percent humidity and 100-degree days they would happily lounge right out in the full sun rather than get back in under the shade. Clearly turkeys can handle heat much better than chickens.

COLD. Again, you would think that a bird that able to handle heat would collapse in the cold. Wrong. We take these birds right up to Thanksgiving, through low teens, and they do fine. They do not gain quite as fast as they do when the temperature is warmer, simply because they convert calories to maintain body temperature rather than add muscle, but they certainly stay healthy and content down to the lower teens. Of course, the waterers freeze at that point so we use those indestructible black rubber 12-inch tubs for water. The 4-inch tubs are okay, but the birds step in them. The taller ones stay clean.

HOUSING. We use the same 2-foot chicken pens for pasture and although the birds can be hard on some of the small bracing, the pens work fine. Although they can't completely stretch their necks as high as they can go, they do not rub the top under normal movement conditions. A turkey neck is quite crooked, or swooped, and really does not come much higher than the top of the bird's back. They move much easier than chickens, cramming up against the lead edge to get the freshest morsels.

We did try, the first year with Tai, putting the birds in an electrified netting yard, but they would not stay in. In fact, we found out that the Premier company recommends its products for everything except turkeys. The birds would stick their necks through and graze contentedly while they received a violent jolt. So much for brains. We also tried a portable house in the field, free range, hoping the birds would come back to it at night. Some came to the house, others went elsewhere and some followed us home. None went inside — they roosted up on the roof.

One fellow in Ontario told me he gets along all right with a roost somewhat like a glorified drying rack. He keeps water and grit there and lets the turkeys follow the chicken pens, scavenging dropped feed. The portable roost keeps the birds protected from predators. Our experience is that the birds end up roosting on the chicken pens and caving in the poultry netting. Oh well, you never get finished refining.

I did meet a fascinating fellow in Alberta a couple of years ago who had built a 50 foot by 100 foot portable turkey corral out of sucker rod. In oil country, this sucker rod, which is a malleable 1-inch square tubing and rolled up on huge reels, is free for the taking. He made 12-foot gates and fastened them together with a couple links of chain so the whole thing would conform to the ground like a big inch worm. I do not know if he skidded it or if it had wheels, but the breakthrough was the segmented long dimensions, allowing it to conform to the ground. He said he didn't have any trouble with coyotes and it worked great. We hope to do some designing on such an idea. I think it has merit, especially since turkeys seem to enjoy walking so much. Giving them a huge area would be helpful and being able to move more at a time would be efficient. The problem is that you have to move it with something other than human labor.

But think about this: what if you built a 10 foot by 10 foot stable for an ox or two, or maybe a couple of milk cows, with three sides tight and one end slatted? The animal(s) grazes through the slats on the one end and the whole stable is mounted on wheels. As the animal grazes, it just pushes the stable forward. Then you hitch it to this turkey corral and you have a portable facility moved multiple times each day without any human intervention except planning. Like I said, we never get done refining.

For now, we're quite pleased with just getting extended use out of our existing pens — we didn't have to build anything new. It just added another $700 gross volume per pen when the pens sat idle anyway (October-November). Not a bad use of resources.

CASH FLOW. The single biggest drawback to turkeys is the longer time between investment and payback. We dress turkeys at 16 weeks, as hatched. Some folks want 20-pounders and others want 12-pounders. By raising both toms and hens we can let one batch meet both desires since the different sexes spreads the weight differences. These 16-week birds average 16 pounds or better. The poults are more expensive as well, running around $1.50 apiece. You have a lot more money tied up longer, but the net return to labor is the same.

MORTALITY. Except for the higher brooding temperature requirements, we have not found any difference between turkeys and chickens early on. The big differences come after about week five. Once turkeys get outside and hit 6-7 weeks, they are almost indestructible. Rather than becoming more and more lethargic, they become more and more aggressive. They seem to blossom, and just get healthier as they get older. Adult mortality simply does not exist. From that standpoint, they are pure joy to raise.

PROCESSING. Turkeys are much more efficient to process than chickens because for roughly the same procedure you have 3-4 times the volume of meat, and the price is better, making the dollar return per hour way higher than broilers. We use four turkey cones for killing, running batches of two. This fall we did 50 birds per hour — at an average 20 pounds (because of a hatchery glitch, we had 18-week-old Thanksgiving birds) that amounted to 1,000 pounds of poultry per hour. Our new 1959 model state-of-the-art rotary scalder handled two at a time just fine, but the picker stalled a couple of times. We had to keep pulling out the biggest feathers, which did not want to drop down out of the picker drum. I explain the main processing difference as one between playing with Leggos and Duplos. The processing advantage more than offsets the production disadvantages.

MARKETING. Here a huge difference between turkeys and chickens exists. Turkeys are very much a seasonal thing — unless you get into turkey ham, turkey salami, turkey sausage, turkey hot dogs, etc. In fact, roughly 75 percent of all the turkey produced in the U.S. now goes through further processing into these other products. Eating a whole turkey is just not something the average person does except at Thanksgiving and Christmas. We sell a few along in the summer, but most go fresh during the week of Thanksgiving. Many people get another one to freeze for Christmas. This is fine with us because it is when we have extra, unused housing anyway. I can tell you that fresh pastured turkey at Thanksgiving is a hot item. All we did was add it to our spring customer letter and order blank. Immediately we had another $10,000 enterprise.

PRICING. Turkey is much more price-forgiving. The reason is that since it is a seasonal thing, people are not so much buying a staple, like chicken, as they are buying a nostalgic, entertainment, celebration vehicle. The special holiday turkey mystique is real, and nobody faults you for charging 3 or 4 times market price — after all, that only amounts to $10-$20 for a celebration meal. The perception is that that is nothing. Chicken, however, enjoys no mystique. It is a staple, and therefore hits price prejudice much sooner. Turkey is a special bird for a special day and folks are more than happy to add a little extra from their wallet to pad the festive occasion.

DIVERSITY. This is not so much a comparison as a point to think about. Since we sell beef, pork, rabbit, eggs, firewood and chicken, we find that each time a customer visits the farm they buy more than the single item that drew them here. Before turkeys, we finished the year with broilers in late September and then beef and pork in late October. That was the last big farm visit until the following May. Now we get more of our customers back one more time during a festive (read that "what else can I buy?") time of year and this stimulates sales of other things, including crafts or woodworking items we may have. Rachel's handmade potholders sell great at this time of year. This additional farm visit can add significantly more than the face value of the turkeys to the farm's annual income.

STACKING. Again, this is not a comparison but a point. Since we can run the turkeys over the same ground as the broilers, it gives us another high dollar per acre item without buying another acre. At 20 turkeys per pen netting $15 per bird, that's $300 net per pen. Moving them twice per day, that's covering 240 square feet per day over an 8 week period (early weeks are only daily moves) for a total of 56 x 240 sq. ft. or 13,440 sq. ft. per pen. That's roughly a third of an acre, which means we're netting $900 per acre on top of the $1,200 per acre with broilers. Not bad for not requiring any more housing, machinery or land. If you don't run them on the same acreage, it means you cover more ground with fertilizer. Either way, you win.

All in all, we are extremely excited about the future for turkeys and plan to expand it as much as is feasible. I think for anyone already doing pastured chickens, turkeys are a perfect additional enterprise, another profit center from your current resources.

Index

Biodynamics 134
Blood, clotted 10
Blood meal 54 *See also Animal byproducts*
Bone meal 54, 323 *See also Animal byproducts*
Boredom 48
Breed. *See also Cornish Cross*
 choosing 31-34
 dual purpose 31
 specialized 31-32
 standard 31-34
Breeder hens
 old 186
Breeding hens
 variables in 158
Brewer's yeast 150, 207
Brittleness
 of bones 241
Broilers. *See Cornish Cross*
Brooder 35-43
 cleaning of 37, 153-154
 makeshift 151
 troughs as 42
 turkey 347-348
Brunetti, Jerry 56
Building ordinances 29
Byproducts, meat *See Animal byproducts*

C

Calcium 57, 205
 deficiency 188
Cannibalism 48, 54
Capping, of bedding *See Bedding, deep*
Carbon-nitrogen ratio 9
 in bedding 37
Carbonaceous material
 as floor covering 36
 in compost 132-135
Carcass weight 16-17, 185, 187, 191
Cardiovascular disease 15
Carotene, and yolk color 344-345
Carpel tunnel syndrome 11
Carson, Rachel 11

Crippling 188, 205-210
Crowding 145, 180-181, 261. *See also Stress.*
Culled hens 21, 144, 269
"Curb Market" 21-22
Curled toe disease 149-150, 206-210
Curly toe syndrome. *See Curled toe disease*
Customers. *See also Consumers*
  bad 228-229, 237-238
  disgruntled 235-238
  environmentally sensitive 56
  no-show 213
Cycles of nature 201

D

Dampness 49
  in brooder 35, 205-206
  in hoophouse 339
Davis, Adelle 206
Death. *See Mortality*
Debeaking 48
Decentralization 8
Deep litter. *See Bedding, composting*
Deficiency 7, 48
  calcium 188
  riboflavin 149, 206-210
Delivery 217-218
Demonstrations 232-233
Detoxification 13, 187
Disease 153-160, 191, 261
  curled toe disease 206-210
  Marek's 208-210
Dogs
  and chickens 181-182
  as predators 165
Dolly
  for moving pens 192-193
Drafts
  in brooder house 155
  in pen 176
Drug resistance 9. *See also Antiobiotics; Resistance*
Ducks 274-277

E

Earthworms 332
Economies of scale 217-218
Education. *See Consumer: education of*
Egg production 198, 256-269
Eggmobile 262-269, 316, 325, 336-337, 339
Eggs 256-269, 325-338
  as protein supplement 324, 345
  quality of 331-332, 343-345
Employees 226-227. *See also Labor*
Environmental sensitivity 9, 56
Equipment, Processing. *See Processing equipment*
Escherichia coli 208-209
Evisceration. *See also Processing*
  mechanical 10
Exercise 15, 46
Exotics 274-277

F

Factory farming 8-9, 54-55
Failure, of business 223-224
Farmer's market 21-22
FASTRACK 52, 58-59, 321 *See also Probiotics*
Fat
  analysis of 15, 318
  in diet 13-17
  saturated 3, 197, 256
  supplement in ration 53
  yellow 15
Feather picker 21, 28, 145
Fecal contamination 8-10, 157
Fecal dust. *See Fecal contamination*
Feed 27 *See also Ration*
  dustiness of 55
  organic 239-248
Feed consumption 20, 256, 262-265
Feed conversion 185
Feed troughs 44-45
Feeders 40
  trough vs. round 45
Feet. *See also Curled toe disease*
  crooked 47

H

I

Insurance
  liability 235-236
Investment 25, 28. *See also Processing Equipment*
  initial 224
  machinery 4
Irradiation 10, 16

K

Kelp 57-59
Kelp meal 52, 57-58
Killing
  electric 194

L

Labor 199, 332-338. *See also Employees*
Lactobacillus acidophilus 58, 205
Lameness 205-210
Laying hens 256-269, 324-338, 339-346
Learning curve 144-152
Leg traps. *See Traps: leg*
Legs
  color of 155
  spindly 155, 315
Let's Eat Right to Keep Fit 206
Liability 235-238
Lifestyle 226-227
  healthy 240-241
Light
  artificial 35
  for turkeys 349-350
Limestone, feed grade 52, 57, 59
Litter. *See also Bedding*
  composting 9
Liver
  for leg problems 149, 315
Livers
  damaged 37
Loose housing
  for laying hens 259-261
Low energy food 46
Loyalty
  customer 213, 224-226, 231, 236-238

M

N

"No-show" customers 213
Noise 182
Nonhybrid. *See Breed*
Nutri-Balancer, Fertrell 56-57, 59, 321-324
Nutri-Carb 321
Nutrient overload 7-8
Nutrient shortfall 7
Nutrients
  from manure 199
  in chicken meat 10
Nutrition 13-15
  inadequate 157
Nutritional deficiencies. *See Deficiency*

O

Oats, crimped 52, 54-55, 324
Observation
  of chicks 48
Open fields
  and predators 163
Opossums
  as predators 166-168
Opportunity
  pastured poultry 2-6, 352
Organ meats
  as riboflavin source 206-207
Organic 239-248, 318
Overpopulation
  of predators 163
Owls
  as predators 169

P

Packaging 217-218
Paralysis 149, 206-210. *See also Curled toe disease*
Parasites
  of turkeys 271
Partitioning
  in brooder 41

V

W

Y

ADDENDUM FOR THE 2010 REPRINTING

The original homemade Pastured Poultry Manual came out in 1991, nearly 20 years ago. Hard to believe we've been at this that long. Of course, we had been raising pastured poultry for nearly 10 years prior to the manual coming out. And I had raised pastured poultry throughout my teen years. Even though we have all these decades of experience, we continue tweaking and developing new things. So much to learn.

I have opted to use the addendum format rather than rewriting portions in the body of the book for two reasons:

1. It's logistically easier.

2. I think it's important for everyone reading this to appreciate how we never quit learning. All you first-timers out there, don't be discouraged. And all you experienced producers, don't get too content in your rut. The day we quit learning and innovating is the day we become worthless.

And remember this: the old saying "anything worth doing is worth doing right" is dead wrong. The truth is that anything worth doing is worth doing poorly first. Nobody gets it right the first time.

Now to detail some of the changes since last printing.

CUT-UPS

One of the biggest changes I've seen over my lifetime as a pastured poultry producer and direct marketer is degenerating domestic culinary skills among the general population. When I started selling chicken in the 1970s, every woman in America knew how to cut up a chicken. Now half of them don't even know a chicken has bones.

If it's not a boneless skinless breast, they don't know what to do with it. Teresa and I recently gave some sausage, eggs, a broiler and some ground beef as a thank you to our banker who had helped us find a bookkeeping error. She called back a day later and asked what to do with a whole chicken. She'd never dealt with one before. Even though she'd raised a couple of children to adulthood, she'd never met a whole chicken.

If you tell the average American that a breaded McNugget shaped like Deno the dinosaur is not a muscle group on a chicken, they don't believe you. As a culture, domestic culinary arts are becoming a thing of the past. While I certainly don't embrace every new technology or cultural fad, I also don't want to become anachronistic. As more and more potential customers complained that we didn't have cut-up chicken, we realized we were missing out on sales. A great number of people simply refused to buy a whole broiler. A whole chicken is intimidating: it actually looks like a bird.

Reluctantly and with much fear and trepidation, we decided to pursue whole bird cut-ups and what we call parts and pieces. The trick, of course, is to charge enough for the value added product to cover the additional labor involved. We know our return to labor on the whole broilers is around $25 per hour, which we consider a professional salary. After all, we are professionals. That's the benchmark. Of course, if the government didn't take half of it in taxes, we could work a lot cheaper, but that's a discussion for a different book.

Obviously we didn't want to compromise that benchmark. To start the process, we cut up ten chickens in a classic approach:

- two drumsticks
- two thighs
- two wings
- one back broken in half
- breast split or quartered, depending on size
- one neck

By timing ourselves, we could establish the additional value per pound needed to return our labor benchmark. Of course, cutting up is not all that's necessary. It must include the packaging, which is a rectangular tray stuffed into a bag. Our price today is $3.25 per pound for a whole broiler and whole bird cut-up is $4.30 per pound. At an average of 4-plus pounds apiece, that's an additional $4.20 per bird.

The key to efficiently doing this cutting is a sharp, heavy, high quality cleaver and a cutting board at least one inch thick. One person cutting and another putting the pieces on the tray and into the bag moves the process along.

Now to parts and pieces. This is tricky because as soon as you start inventorying pieces, they have to sell equally. You can't end the season with 2,000 pounds of backs and no boneless breast. To be sure, we consider parts and pieces a luxury item. Anybody can learn to cut up a chicken. But for those who don't or won't, we now offer that service at a luxury price and don't apologize for it.

First, we weighed ten chickens and then cut them into parts. Then we weighed all the parts. Then we started playing with numbers, assuming that the tenders would be the most expensive. Actually, we went to the supermarket to get the industry price ratio. If the industry charged twice as much for a boneless skinless breast as for a drumstick, we wanted to know that.

In order to not belabor this discussion, let me cut to the chase. Here is our parts and pieces price per pound to get an equivalent return to labor as a $3.25 per pound whole bird:

Boneless skinless breast, 2 per pkg., avg. 1 lb.	$13.00
Tenders, 6 per pkg., avg. 1 lb.	$14.00
Legs and thighs (kept together), 2 per pkg., 1.5-2 lb.	$ 4.50
Wings, 8 per pkg., 1-2 lb.	$ 3.25
Backs for stock, 3 per pkg., 5 lb.	$ 1.00
Neck for stock, 6 per pkg., 1-2 lb.	$ 1.00
Feet for stock, 30 per pkg., 3 lb.	$ 1.00
Hearts and livers, 8 per pkg., 1-1.5 lb.	$ 3.00

The bottom line here is that a customer can buy a whole chicken for the same price as the breast. One of the reasons these prices are high is because we do this by hand rather than by machine. We don't do enough volume to pay for a multi-million dollar machine like the industry uses. We also don't turn people into machines by giving them repetitive motion disorder tasks for eight hours a day.

Profitability in all of this cutting up depends on larger birds. The meat-to-bone ratio of a 5 pound bird is dramatically different than it is on a 3.5 pound bird. When we do parts and pieces, we separate out the birds exceeding 5 pounds. That makes the breast pieces, for example, large enough to justify the cutting procedure to extract them. Dave Schaefer of *Featherman* fame shot a couple of instructional videos here at Polyface to teach people our cutting procedures.

Like all skills, technique is everything. We use ice beds during parts and pieces to keep things cold until we get them in a package. We used to put these parts on trays like the whole bird cut-ups, but eventually discontinued that since virtually all of our parts and pieces are sold frozen. Customers don't paw through the stash; they just grab one and go. Or we put it in their delivery bag and they see it when they defrost it. That saves lots of packaging, time, and money.

Our label for all these products is a generic label. It lists all the possibilities, including stewing hens, turkeys, whole birds, and parts. We use a sharpie to darken the blank in front of the particular item in the package. No matter how hard you try, you can never discern what part or piece is in the bag once it's frozen. It all looks the same. Always put on a label and mark it, not just for the customer, but for your own sanity. We get our labels made at the same place that makes our federal inspected labels for beef and pork. They use a special kind of glue that regular print shops don't use. If you put on a label, you want it to stay on.

Every year we do more and more further processing. The first year we offered this service, we made $20,000 more without raising a single additional broiler. That's the power of value adding. And we made lots of customers happy. Happy customers tend to be more loyal and try other things.

One rule of thumb about pricing: if 10 percent of your customers are not complaining about price, you're too low. And if you think you don't deserve a professional salary, you might not have what it takes to produce a professional chicken. Farmers will become respected again when we respect ourselves enough to think and act like professionals. Never apologize for being a professional and receiving a professional salary. In the final analysis, is a great farmer less valuable than a great baseball player, a great actress, or a great singer? That's a rhetorical question, by the way. The answer is supposed to be obvious.

SOYBEANS

Largely through the writings of the Weston A. Price Foundation (WAPF), many people are concerned about chickens fed soy products. This is a ticklish subject for me because few people are bigger fans of Sally Fallon and WAPF than I am. The argument revolves around estrogens and dangers from soy derivatives.

Ultimately, the answer is probably that people should eat less chicken. Historically, common people didn't eat chicken; only royalty ate poultry. As an omnivore, chickens can't survive on forage like herbivores can. Traditionally, chickens scavenged a few table scraps. They were not a primary enterprise because grain was too precious. When grain required hand planting, scything, shocking, hand winnowing and flailing, shoveling, and storage in crude wooden bins susceptible to rat infestations, it was a precious commodity. Having enough grain leftover after making flour for human consumption was simply not an option except for royalty. Hence, poultry and kings.

People who want to eat chicken a couple of times a week are certainly not traditionally pure. So let's all be honest about compromises and what is not natural. It's not natural to eat chicken more than once or twice a year. Furthermore, double breasted chicken is not natural. To grow these very young, tender birds like western cultures have come to expect requires a different diet than the one provided scrawny scavengers around the old country homestead. These high octane race car chickens can't survive on scavenging.

Teresa's grandmother said that when she was a little girl in the 1920s, if they wanted chicken for the Fourth of July, they had to set a hen on eggs the first of January. That chick would hatch before February 1, then it would grow for five months until it was big enough to eat. Today, with high octane feed and genetic selection (not genetic modification) broilers get just as big in about five weeks. Plentiful grain (diet) and genetic selection account for the difference.

That being said, let's look at the alternative dietary items. Soybeans provide protein. Here are some substitutes: earthworms, mealy bugs, fish meal, winter peas. Earthworms are fine for a backyard operation, but are a logistical nightmare for a commercial outfit. To offer 1,000 pounds of earthworms a day would require a massive earthworm operation, which itself would have to be fed something. An earthworm farm of that scale probably doesn't fit a natural system either. And even if you could raise that many earthworms, handling them and feeding them to the chickens is problematic. They are squishy, perishable, and hard to mix into a feed ration.

How about mealy bugs? Many of the same problems as earthworms. To grow and farm that many bugs—any bugs—in concentrations big enough to make a dent in the protein requirements of a commercial poultry operation exceeds natural carrying capacity. It becomes the proverbial tail that wags the dog.

Fish meal? Yes, that's possible, but the menhaden used for this is the feed source for carnivorous ocean fish. The oceans are already being mined beyond sustainability. And many of the fish sources are the result of Japanese drift nets that indiscriminately catch everything and epitomize anti-ecological practices. Do we really want to eat chickens that represent the demise of the oceans? Probably not a good tradeoff.

Winter peas have some potential, but they thrive only in small areas. As an experiment, this year we are raising a batch of broilers on winter peas, but we are shipping them in from Montana. We buy non-GMO (Genetically Modified Organism) soybeans from neighbors right here in Augusta County, Virginia. In the big scheme of things, is it better to use local GMO-free soybeans or import peas from 2,000 miles away in order to be a purist? And what kind of purity is that if we consider transparent commerce, diesel fuel, and local economy?

The research impugning soybeans, to my knowledge, has not been done on the whole bean, but only on derivatives, like soybean oil, or soybean meal (the oil is pressed out) or some other portion of the soybean. To demonize the soybean for bad affects of its parts and pieces is like demonizing corn on the cob because high fructose corn syrup causes diabetes. We all know that whole foods have natural checks and balances that do not exist in processed food. I'm not ready to throw out a whole food as being inherently evil.

In our ration, we use only whole roasted soybeans. Not parts and pieces. They taste like peanut butter. I figure anything that tastes like peanut butter can't be too bad. Using the whole bean preserves the balances and holds onto the oil and enzymes in their natural form.

Beyond that, we depend on the grass to cure anything that's not right. Over the years, Polyface pastured chicken has been subjected to countless empirical and esoteric tests: chromatography (biodynamics), pendulums (dousing), nutrient analysis, and frequency vibrations. Each time, this chicken passes with flying colors. Why? I think it's the grass. It's all about the grass. Chlorophyll is nature's number one detox agent. Whatever isn't quite right, chlorophyll will heal and make it right. That is why managing pastured poultry for maximum salad bar intake is the foundation for vibrant, healthy chicken.

TURKEYS

I think we've cracked the turkey code. For the last couple of years we've had great success with turkeys after suffering years of withering defeats. Turkeys seem to have one goal in life: to connive a more creative way to die. After watching us suffer through this for a couple of years, my daughter-in-law, Sheri, finally said: "I didn't know raising turkeys was hard, 'til I came to Polyface." Ouch.

On her family's farmette outside Austin, Texas, turkeys brooded with chicks and did just fine, thank you very much. I protested: "But the books all say you can't put turkeys and chickens together."

"And since when did you start believing everything you read in books?" she laughed. Got me. We'd tried everything else. One year only one survived out of 300. What did we have to lose?

We began brooding them together and have never looked back. Last year we raised nearly 1,500 and this year plan to raise more than 2,000. The market is wonderful and turkeys are magic on pasture. I'll start with brooding and go straight through the pertinent protocol at our current level of success.

The magic number for brooding is one turkey poult to five chicks. We don't change the brooder temperature for the poults (normally poults like it five degrees Fahrenheit warmer than chicks). The only thing we change is to put in a slatted box with waterers on it because the turkeys are slower at discovering the water nipples. The slatted box keeps the birds from kicking bedding up into the waterers.

At the one-to-five ratio, the turkeys bully their way into the warmest spots, hogging the heat they so desperately need. They also shove their way up to the feed first and pick out the high protein particles. Poults typically want nearly 10 percent more protein than chicks. But rather than change the ration, we use our regular broiler ration. The poults select what they want since they are bigger, but not enough to deprive the chicks what they need.

After a week, the turkeys find the water nipples and we remove the waterers. The chicks teach the turkeys about feed, water, bedding, and life. They conduct seminars on marriage and adolescent behavior. The point is that as long as you maintain five chick teachers to one turkey student, you will have educational and social success.

At or before three weeks, the chicks and poults go out into the broiler shelters—at roughly the same ratio. By week seven, the turkeys become obnoxious to the chicks and we remove them from the field shelters. Of course, that gives the chicks more room for their last week or two.

The turkeys go into a feathernet—the electrified poultry netting. We use Premier's brand. At that point, the turkeys are big enough to not fit through the holes but small enough to not bowl it over. Unlike every other animal, the turkeys never seem to figure out the electrified netting. Even when fully grown, they stick their necks through the holes to grab a grasshopper or cricket on the outside. You can watch them jerking and twitching as they get shocked, their long necks rubbing the netting. But they just look around nonchalantly as if electrocution is a wonderful way to live.

I think it's critical to expose the poults to the netting while they are still small enough to be physically restrained by it. If you wait to expose them at twelve weeks, they are big enough to walk over it and don't learn to respect it as a physical border to their world. Once they learn that their world has an enclosure, they don't challenge it even when they are fully grown and can practically look over it from a standing position.

Turkeys need copious amounts of grit. Five hundred twelve-week-old birds will eat nearly 100 pounds of grit per day. Grit comes in various sizes, and as the birds grow, we increase the grit size. We keep it in front of them, all they want. The last month, the stones are about as big as marbles, but the birds ingest them by the 50-lb. bag. I think some of the land healing magic with turkeys is this mineral, biologically activated inside the bird, and then spread on the land. This silica makes the subsequent grass shine like something out of a fantasy world.

Because turkeys require such minimal shelter, they are compatible with steep slopes and rugged terrain—places you wouldn't want to pull broiler shelters. The netting can go up and down any area and the turkeys certainly don't mind rough ground. They seem to thrive on it. They don't even need impermeable shelter. Nursery shade cloth is fine. If it rains, they just get wet but that's fine. Also as they grow bigger, they are less vulnerable to predation. All in all, we are bullish on turkeys and hope this helps make it easier for you.

THE SCISSOR TRUSS

I'm a fan of hoop houses. But when it comes to portable structures for poultry, I think the scissor truss beats the hoop house.

The foundation is the same: pipe skids. But rather than attaching hoops to the skids, we weld an angle iron ear that can receive a 2" X 6". Forming a squat X with these main timbers, we put an A-frame over the top of the squatty X, with some appropriate vertical and horizontal bracing, and (voila!) we have an incredibly strong structure that utilizes vertical space.

Inability to utilize vertical space is a deficiency of the hoop house. Yes, you can attach boards to the hoops, but the board ends must be inside the plastic. That means all the attachments are right at the end of the boards, which encourages splitting. Since we have our own lumber (woodlot and band saw mill) we find the scissor-truss and A-frame type roof a more compelling design.

That also gets the peak up high enough to create a catwalk above the intersection of the X-truss with enough head room to walk comfortably. The nest boxes, which hang on either side of the catwalk, are high off the ground. This does two things. First, it reduces chicken loitering, which reduces breakage. Secondly, it creates a foot-drying and cleansing route from the ground to the nest boxes. That time and space keep the eggs cleaner.

Perch boards on the trusses provide adequate lounge space for the birds. Although this design is a bit heavier than the hoop house, it's not too heavy and the additional weight protects the portable structure from wind movement.

The most important reason, however, for this new design is that it creates structural integrity higher off the ground. That means nothing exists at ground level to catch legs or squish chickens when the structure is skidded to the next spot. Ground level bracing on the hoop houses creates logistical issues for moving because wayward birds can get caught. The hoop house needs ground level bracing to keep the hoops from spreading and/or shifting into a parallelogram. If the bracing moves just two feet off the ground, then it's a nightmare to navigate when you're walking inside.

The scissor-truss has nothing on the ground except the two skid pipes, which are moving in a straight forward direction and therefore not liable to snag a bird. Only sideways movement is dangerous for the birds. And the bracing doesn't matter because the only place you have to walk is through the catwalk to gather eggs.

If birds stay inside the scissor-trussed design, it simply moves over them while they stand there. With the hoop houses, ground-level bracing or framing can easily catch the wayward bird and squish her; or the move requires another person or two, in addition to the tractor driver, to shoo the birds away from bracing and framing. This major difference makes moving much simpler, safer for the birds, and more efficient. And because it fully utilizes vertical space, the total structural footprint for adequate shelter drops by a third for the same number of chickens. And the chickens are happier because they enjoy more off-ground resting. We call this structure the Millennium Feathernet.

GUARD DOGS

Four years ago, after losing several nights' sleep over a predation issue, we purchased a guard dog trained by a Maryland breeder to bond with chickens. It was one of the best decisions we've ever made. As a general rule, these days we don't let chickens go the field without a guard dog present.

I don't think the breed is necessarily as important as the training. We've found that puppies are playful and harder to break from recreationally enjoying the chickens than older dogs. I guess they're like people: they kind of mellow out as they age.

For training, the best investment we've found is a Cabela's buzz collar. This is a remote-controlled collar full of metal beads. When the dog does something you don't like, you just zap him and the beads buzz, giving the sensation of an electric shock. Although some breeders say these dogs have a hair coat too thick for these to work, we have found them extremely effective, even on thick-haired dogs.

The beauty of these collars is that the trainer doesn't have to be present to administer the discipline. The dog gets zapped for misbehaving and assumes it's the hand of God. Far more effective than having to be present. That way too, you can continue to be the dog's friend and he never understands yours was the hand of correction.

Another thing we've done that breeders tend to say won't work is to train these dogs to two strands of electric fence. While it is a bit of work to add a second strand of fencing at about the 12-inch level, it's not nearly as much work as gathering up 50 dead 8-week old broilers after a predator attack. Or what happened to us one year when over the course of two weeks a pair of foxes with a den of babies began pilfering hens out of the eggmobile. Because they carried them off to their den, we didn't realize anything was going on until we noticed the chicken numbers were low. That's why people put chickens in factory houses.

I think it's interesting that many of our customers patronize radical environmental organizations that revere the very critters that make pastured poultry difficult and expensive. The right hand doesn't have a clue what's going on with the left hand. As pastured poultry re-injects humans and livestock into more natural conditions, it requires re-balancing wildlife. As synthetic fibers and animal worship have destroyed the live fur trade, the proliferation of furry chicken killers is astronomical. This includes hawks. Now that a hawk perches on nearly every fencepost along the interstate, can we finally take them off the endangered species list? Come on, people, get real.

A guardian dog is a great equalizer. If a dog kills these predators, somehow people don't mind as much. If it's a bullet or trap, shame on the farmer. But a dog is just expressing its dogness, so that's okay.

One way to help train the dogs to the electric fence is to put a chain-type dog collar on them and let them drag 18 inches of the chain on the ground. That way if they come up to the fence and put their head over, the chain hits the wire and shocks them. It also slows them down a bit during a full-tilt run.

384

Because a dog can only patrol about 20 acres, we move the dog from field to field every few days. The residual, or halo effect, gives effective coverage over far more acreage than the dog can physically cover at once. Once the dog is removed from a field, it takes the wildlife a few days to realize he's gone. That's the halo effect. The dog enjoys being moved around too, receiving a change of scenery. Using this method, we've been able to increase the effective coverage to about 60 acres with one dog. It's not 100 percent, but close.

Although these dogs are expensive, their cost is nothing compared to a good night's sleep and no predator attacks. We recommend nocturnal guardian dogs to everyone who has a substantial volume of pastured poultry.

HERITAGE BREEDS

Here at Polyface, we use non-hybrid dual-purpose breeds for our laying flock. Right now, we're using Rhode Island Reds, Barred Plymouth Rocks, and Black Australorps. Those are traditional breeds that would have been found on any American homestead a century ago. These don't lay quite as well as the hybrids like Dekalb Golden, Golden Comets, Cherry Eggers, J.J. Warren Cross and Sex Links—all crossed with Leghorns.

But the hybrids have a couple of shortcomings. First, they lay too prolifically for their metabolism to keep up. These smaller-bodied birds, laying six eggs a week, equivocate to a 150 pound person losing 15 pounds a day. Can you imagine trying to keep up with that regimen? This lay rate does a couple of things. First, the bird begins cannibalizing its own skeleton to keep up with egg shell calcium and other egg nutrients. Second, the birds can't ingest enough green material to make the dark, rich yolks characteristic of nutrient-dense eggs.

Their total energy requirements are so high, they can't sacrifice digestive energy on something as low in energy as grass and clover. They want corn. The non-hybrids like the ones we use drop down to five eggs a week and their bodies are half again as large (5 pounds compared to 3 pounds). Their production would equivocate to that 150 pound person losing 5 pounds a day. While that is still high, it's doable. Although I haven't seen scientific studies to back this up, intuitively it seems to me that a bird that's not cannibalizing her body to keep up with production will pack a little more nutritional punch in her eggs. That just seems reasonable.

Another characteristic difference we've noticed is that the bigger, traditional birds are less flighty. That means they stay in the feathernet (electric fencing) better. They don't go as crazy if they are frightened. They are smarter—they actually watch for hawks. They are hardier—they handle hot, cold, rain and wind better. It stands to reason that they wouldn't be as fragile as the smaller, higher strung birds.

Polyface eggs have a dominant reputation in the mid-Atlantic area. I think one reason is that we have not succumbed to the lure of hybrid egg laying genetics. Yes, it means our production per bird is lower by about 25 eggs per hen per year. Yes, it means our birds eat a little more grain per egg. But, it also means our birds live longer, stay healthier, eat more forage as a ratio of total diet intake, and yield a wonderful carcass as a stewing hen. Most pastured egg operations are using the hybrids. Some even use debeaked birds, which are at a distinct disadvantage for eating bugs and forage.

One of the problems we're having at Polyface is that the nationwide breeding stock flocks for these minor breeds is so small that when we order 3,000 pullets from a hatchery, they can't come in one group. They have to be spread out over a month. That's a nightmare because instead of starting all the chicks at once in the brooder, they come into production on a staggered schedule. They grow at different rates. They are ready to go to the field at different times. They begin laying at different times. Instead of all starting at once, they dribble in over a long time.

This is one reason I'm encouraging farmers to use these minor breeds, because that's the only way the breeding flocks will increase to accommodate larger commercial numbers. When a customer buys a Polyface egg, the ripple effect is huge. That egg keeps a small pasture-based farm in business. It puts a minor breed bird in the field—a smarter, minor breed bird, mind you. It creates a marketable stewing hen at the end of productive life. It patronizes an independent hatchery, which is key in preserving non-industrial poultry. It creates market demand for traditional genetics and encourages existing seedstock flocks to stay in business. That's a pretty cool ripple, don't you think?

As an aside, let me address the radicals in the animal rights movement who keep lobbying and trying to criminalize the shipping of poultry as inhumane. Birds are not mammals. Chicks are fine for up to 72 hours without food and water. As soon as they eat or drink, however, they need to eat and drink several times a day. Ideally, 48 hours is the cut-off, but they will hang in there up to 72.

I was speaking at the Washington D.C. Live-Green expo recently and during the Q/A period a lady asked if I endorsed Murray-McMurray shipping chicks around the country. Incidentally, I picked up some brochures from one of these groups and it mentioned Polyface by name as a hypocrite. The section about us could not have been any more wrong. It was total untruth, written by someone with an ax to grind—some hollow-eyed vegan no doubt who couldn't keep up with me doing meaningful work even for an hour.

I responded that decades ago, when we started, we purchased chicks from a local hatchery. Then it closed. Then we purchased from a farther hatchery. Then it closed. Now we get them out of Ohio or Texas because we can't get them in Virginia. My dream would be to get them from Virginia again, but we need a hundred Polyfaces to create the market demand for that infrastructure to return. Unless and until it does, however, these independent hatcheries and air freight shipping are absolutely the lifeline for the non-industrial poultry movement.

She walked out before I even answered the question. She should have been more honest and asked: "How can you eat chicken when we all know a chicken is the same as a human child?" These animal rights radicals wanting to shut down chick shipping (you have to say that carefully in order to not have a slip of the tongue) are not trying to create chick welfare. They think it's sinful to eat chickens. What's ironic is that they duplicitously fall into a line of thinking that the industrial poultry movement loves: annihilate the small-scale poultry competition. Trust me, the industrial poultry folks love these animal rightists trying to shut down chick shipping. If these radicals succeed, it will destroy the final alternative to Tyson. And wouldn't Tyson love that?

So while they say they are doing this for the chicks, they are actually playing right into a favorite agenda item of the industry: get rid of the pesky independent producers. Before you unleash righteous indignation on something, you'd better be sure about the world such passion will create. The road to hell is paved with good intentions. Righteous indignation is a powerful force. Release it wisely and judiciously, or you might create a worse world than the one you're living in now.

Believe me, Tyson loves it when these animal rightists petition against chick shipping. They probably throw chicken barbeque parties to celebrate their friends at animal rights organizations. True compatriots. Duplicitous do-gooders. Gullible zealots.

Now let's go to the broilers, where Polyface is more vulnerable. There, our farm compromises and uses the industrial double-breasted NASCAR race car high octane broiler. Fast growing, heavy breasts. So far, these chickens haven't gone the way of turkeys, where natural breeding service is no longer doable. Chickens are still enjoying natural service. Whew! But these chickens still grow unbelievably fast and are temperamental as a result. They are prone to leg problems, heart attacks (their muscle grows too fast for their organs to keep up), and respiratory issues.

Why would we use such an unnatural chicken? Marketability. We tried for several years to offer a non-hybrid bird, but people didn't want it. Marketing is persuasion, and in persuasion you can't push people beyond their tolerance level. If a 1 represents a McDonald's junkie and a 10 represents a true blue foodie with grain mill in kitchen, lard making in slow cooker, local everything and scratch everything, you never try to move a 1 to a 10. You try to move a 1 to a 2 or 3. Otherwise, you're just offensive and considered obnoxious.

The world will only let you be so weird. You can be a nudist, and you can be a Buddhist, but a nudist Buddhist—that's too weird. So here we are telling people they should buy their chicken somewhere besides Wal-Mart; they should cook it themselves (that's quite a stretch for many these days); they should pay a little more for it. To go beyond that and tell folks not only that, but you also want dark meat, a razor breast, and a chicken too old and tough to fry—you've just become a nudist Buddhist.

Even people who touted themselves as real heritage aficionados didn't want these non-hybrid birds. We used White Plymouth Rock cockerels, 12 weeks old, and called them Marco Pollos (pronounced in Spanish, poyo)—Old Country birds. Here was the shtick: "You may not have been able to sail with Magellan around the cape. You may not have been with Columbus. But you can smell the smells and see the sights, direct from the galley, with the Marco Pollo." It didn't sell. After three years, we discontinued it; simple as that.

What good does it do to go bankrupt being altruistic? At the end of the day, we need to pay our taxes and keep shoes on our feet. Does that mean I wouldn't love to retry this? Of course I would. These birds were healthier, tastier, better grazers. They just don't have a double breast and the meat is tougher and more of it is dark, due to their additional exercise. These are not lethargic birds; they get up and run around.

Does this compromise mean I've joined the dark side? Am I a hypocrite? Have I joined the enemy? Perhaps. I confess it's all a problem. And I don't have all the answers. If we hadn't tried, I'd be more contrite. But we tried. Unless and until we have more people willing to go farther in their 1 to 10 persuasion, we'll produce a bird light years better than industrial but with the double breast that consumers have come to expect.

Generally, we don't worship breeds. In every breed, you will find animals that thrive under a given set of conditions and ones that won't. The idea is to find the genetic base that works for you. And that will be a completely different base from what will work best in an industrial confinement house.

I would be remiss in this whole genetic discussion if I did not introduce every reader to the work of the American Livestock Breeds Conservancy (ALBC). This organization is dedicated to preserving heritage genetics. The goal is to preserve genetic diversity. I deeply appreciate what ALBC has brought to the table and wish more people would use more of these traditional breeds.

That said, let me finish this whole discussion by going off the lunatic edge and introducing you to linebreeding and what I call nativized genetics. Every generation of a plant or animal carries genetic memory that makes subsequent offspring a little more adaptive to the area's ecology. Any Floridian who roomed with a Canadian in college is aware of this adaptation. The Canadian can walk around in a 60 degree room in short sleeves; the Floridian needs an L.L. Bean parka in the same room.

Heritage breeds were developed over many decades, primarily in Europe, and exported to America. These breeds form the basis of what we now call heritage breeds. They were not developed in America. Many of them carry geographic names, like Scottish Highlander cattle or Yorkshire pigs or Suffolk sheep. Farmers in those regions gradually selected the phenotype that proved functional in that area; hence, the geographic names.

I suggest that we should be doing the same things in America. Rather than just preserve Old Country breeds, we need to be developing our own heritage breeds. I think a Swoope cow would be great for my great-grandchildren to enjoy. Essentially, I'm talking about using the same kind of functional selection process used in Europe to give our future farmers a similar legacy. New breeds. Geographically specific, adaptive, functional phenotypes. How about that?

I confess that I'm not impressed when I meet someone in Alabama exuding about their Scottish Highlander cattle. Why do they have them? Because they're cute. They're different. But these animals were selected for centuries to live in cold, rugged, mountainous conditions. Alabama is practically abusive to them. This is not the way to preserve heritage genetics. They need to be appropriately climatically sited.

So how do we create such a genetic legacy? Linebreeding. Perhaps we could even call it wild breeding. Think about it. Have you ever heard anybody say: "We've got incestuous opportunities out here in the deer population. Goodness, a father might breed his daughter, or a son his mother. We'd better come in here with a helicopter, bring some of those bucks out of there, and move them 50 miles away. Then we'll bring a few bucks from there back here to make sure we have some outcrossing."

No, nobody says that. And yet the wild populations go along just fine with a somewhat haphazard familial breeding program. And they seem to get along fine. And they look amazingly similar. That's because functional phenotype and physiology are the only criteria. Not some arbitrary human standard of size, color, or whatever. Over time, this system creates the kind of consistency that's normal in wildlife. Aberrations are extremely rare, and usually don't last long.

Think of how consistently similar deer or squirrels or zebras look. In a herd of zebras, the adult females don't vary by more than a few pounds. The difference between the biggest one and smallest one is practically imperceptible. That's because they've been self-selecting for functionality for a long time. No sophisticated breeder walked in one day and started selecting the ones with the fuzziest tail or the most pointed ears.

391

Daniel started his rabbits when he was eight years old, as a 4-H project. As of this writing, that's been 20 years ago. For roughly five years, he endured 50 percent mortality. He fed them forage, did not medicate or vaccinate, and did not bring in any additional outside genetics. Close breeding and patience paid off. Gradually, all the maladies began to subside. His rabbits experienced all the problems you see in rabbit rearing books: long teeth, sore hocks, coccidiosis.

But gradually they started to change. Now, 20 years later, without any outside genetics, no medications, no vaccinations, he has the most homogeneous-looking bunch of rabbits I've ever seen. They are cookie-cutter consistent. That's because their appearance is completely functional and not based on anything else. And they haven't been outcrossed, mongrelized, or genetically compromised. Indeed, this is now a Daniel breed, truly nativized.

Fortunately, because rabbits have multiple offspring and a relatively fast generational turnover, Daniel has been able to do this in just 20 years. I figure to do the same thing with our cow herd will take 80 years. Oh well, we're five years into it. But you have to start somewhere.

Some people are calling this wild breeding. Some cattlemen who have adopted this idea are amalgamating their cow herds, running multiple bulls, and not worrying too much about record keeping. If a cow doesn't have a calf, she's culled. They keep bulls from within the herd, from older cows. No cows receive calving assistance. If they die, they die. This hands off approach, while it may sound harsh, or even uncompassionate, is really the most efficacious way to create genetic strength, overall health, and functionality in an environment. Ultimately, I don't have a problem with that.

Well, you say, what about the suffering cow that's trying to calve and finally dies out there? I confess that's not my style, but I completely appreciate the process. Would it have been better to intervene with attention and medication when Daniel was losing 50 percent of his rabbits to birthing problems and sickness? If we had, we would not have the healthy, model group of rabbits we have today. Everyone who sees them exclaims about how healthy and consistent they look. I wish I had a cow herd that looked that way.

The relationship between pain and gain is real: a difficult marital discussion to gain new intimacy; a difficult exam to gain new credentials; a difficult exercise regimen to gain new wellness. You know the Marine slogan: "Pain is weakness leaving the body?" The point is that abundant life is not fluff and fuzzies. Genetic selection is the same way. To really make progress takes some strong culling. That's natural.

What we need in pastured poultry is this kind of nativized genetic selection. Wouldn't it be neat if we selected breeding stock that grazed more aggressively or saw hawks and ran under shelter? If you put a flock of hens in an eggmobile and let them get picked off by predators until only 2 remained, and bred those, and did that over and over for several generations, maybe you could increase predator awareness and survivability. You could call them MacGyver chickens or something.

From a legacy standpoint, I think that would be superior for subsequent generations than breeds developed in Europe a couple of centuries ago, frozen in time and type. By practicing the same kind of functional selection used in the old country to create heritage breeds, we could create additional bioregionally specific hardiness and add genetic diversity to subsequent generations.

This year, 2010, here at Polyface we are trying the Freedom Rangers, a buff colored meat bird developed in France. The genetic stock has now been imported to the U.S. and we'll see how these birds grow. And how customers receive them. As this whole pastured poultry movement grows, I'm convinced that specialty genetics and hatcheries will eventually produce the phenotypes that perform better on grass.

PROCESSING REGULATIONS

As the local cottage-based pastured poultry movement grows, the food police fight back with more vengeance. Many people wonder why I'm more vituperative toward the government bureaucrats than the industry, and the reason is simple: Tyson doesn't employ attorneys general and sheriffs wearing side arms. Tyson can't do anything to me. The government can, and does.

But as monstrous as the food police are, clever paths exist over, around, and under the regulations. I want to share a great success story to encourage people to persevere in the face of adversity.

One of our former apprentices, Tyler Jones from Corvallis, Oregon, received a visit from the food police. He wanted an open air shed facility like we use at Polyface. Simple, cheap, and ultimately sanitary due to the fresh air and sunshine wafting through. But the Oregon food police demanded walls and windows.

Tyler cleverly realized that the regulations did not specify the size of the walls nor the dimensions of the windows. He has a permitted facility with walls 8 inches high. The windows are 6 feet high—sliding patio doors from the hardware store. Essentially he has an open air facility because he can slide open the windows and use the screens. The walls are impervious, just like the regulations call for. The windows have screens. It's all a wonderful example of creative compliance.

The bottom line is this: don't quit. Persevere. Think creatively. Regulations are made to circumvent, so enjoy the challenge. Don't let the food police quash your dreams. Just go for it.

Figure 1. The scissor truss design of the millennium feathernet offers more protection, safety, and better use of vertical space than the old hoop house design for portable skid structures. *(Photo by Rachel Salatin)*

Figure 2. The millennium feathernet uses a catwalk for both human and chicken access to the nest boxes. The A-frame roof design on top of the scissor truss is not only strong, but keeps the birds high and dry. *(Photo by Rachel Salatin)*

Figure 3. The Roostmobile for turkeys uses the scissor-truss design on a hay wagon chassis. The birds can roost on the perch boards and the whole structure is highly mobile. By running the X parallel to the chassis, the whole contraption can be elongated to create more shelter. *(Photo by Rachel Salatin)*

Figure 4. A lightweight hoop trailer using aluminum conduit and canvas can be pulled around by hand or with a simple 4-wheeler (All Terrain Vehicle). It can be moved daily within a two or three day netting setup to spread manure more evenly. *(Photo by Rachel Salatin)*